Springer Series in Information Sciences 31

Editor: Teuvo Kohonen

Springer
Berlin
Heidelberg
New York
Barcelona
Budapest
Hong Kong
London
Milan
Paris
Tokyo

Springer Series in Information Sciences

30 **Self-Organizing Maps**
By T. Kohonen

31 **Music and Schema Theory**
Cognitive Foundations
of Systematic Musicology
By M. Leman

Marc Leman

Music and Schema Theory

Cognitive Foundations of Systematic Musicology

With 101 Figures

Springer

Dr. Marc Leman

University of Ghent,
Institute for Psychoacoustics and Electronic Music,
Blandijnberg 2, B-9000 Ghent, Belgium

Cataloging-in-Publication Data applied for.

Die Deutsche Bibliothek – CIP-Einheitsaufnahme

Leman, Marc:
Music and schema theory : cognitive foundations of systematic musicology / Marc Leman. – Berlin ; Heidelberg ; New York ; Barcelona ; Budapest ; Hong Kong ; London ; Milan ; Paris ; Tokyo : Springer, 1995
(Springer series in information sciences ; 31)
ISBN 3-540-60021-3
NE: GT

ISBN 3-540-60021-3 Springer-Verlag Berlin Heidelberg New York

Printed in Germany

Typesetting: Data conversion by Kurt Mattes, Heidelberg
SPIN: 10480074 54/3144 - 5 4 3 2 1 0 - Printed on acid-free paper

To Jan L. Broeckx

Preface

In 1987, when I started to set up the research facilities at the Institute for Psychoacoustics and Electronic Music (IPEM) of the University of Ghent, Belgium, music cognition was still dominated by a symbol-based paradigm – inspired by computational linguistics. Music was conceived of as a set of symbols (like the notes on a score) on which rules were operating. Being aware of the limitations of this approach, the projects at IPEM have attempted to give music cognition a foundation in sound, rather than scores. New developments in psychoacoustics and, above all, the new and radical methods of the subsymbolic paradigm have been a source of inspiration on which the present approach has been based. This monograph summarizes the results of my research over the years and explores new paths for future work. The aim is to give musicologists, students, researchers and interested laypersons a profound introduction to some fundamental issues in the cognitive foundations of systematic musicology. This is done by means of a case study in tone center perception but the results are extrapolated towards other modalities of music cognition, such as rhythm and timbre perception.

An *interdisciplinary* viewpoint had to be adopted which includes results of musicology, psychology, computer science, brain science, and philosophy. In order to make all this accessible to a general audience, care has been taken to make the text as self-contained as possible. The technical language has been restricted to the most elementary concepts.

The structure of the book is as follows. After a short introduction, Chap. 2 focuses on the problem of tone semantics from a historical point of view. In the second part of this chapter, the main achievements of recent research in music perception are discussed. Chapter 3 is about the decline of the traditional phenomenological approach to pitch perception and introduces more modern ideas on pitch perception by means of a discussion of auditory illusions. Chapter 4 presents a framework for a computer model of music perception. A distinction is made between different types of representations, including images and schemata. The auditory model on which artificial perception relies is discussed in Chap. 5, whereas Chap. 6 introduces the reader to a model of learning by self-organization.

In Chaps. 7–8, it is shown that a schema (or mental knowledge structure) for tone center perception emerges by mere exposure to musical sounds. In

Chaps. 9–10 it is shown that the model for tone center recognition and interpretation can be used as a tool for analysis in musicology. (Applications for interactive computer music are straightforward but are not explored in this book.) Chapter 11 extends the ideas to the domain of rhythm and timbre perception. The last two chapters, Chaps. 12–13, relate the model to neurophysiological foundations, theories of meaning formation, and historical developments in musicology. The final chapter describes the background for a psycho-morphological approach to music research.

This book could not have been written without the help of many colleagues and friends. First of all, I wish to thank H. Sabbe for his continuous support and stimulating ideas and D. Batens for valuable philosophical discussions during the initial stage of this project. Special thanks go to E. Terhardt of the Technical University of München for the use of his auditory model, and to J.-P. Martens and L. Van Immerseel from the University of Ghent for help with the adaptation of their auditory model. F. Carreras from CNUCE/CNR at Pisa ported the SOM implementation to the nCUBE2 and gave many valuable remarks on the final draft. Thanks also to A. Camurri, R. Parncutt for reading the first draft of this book and to N. Cufaro Petroni for helpful suggestions, in particular during the development of the attractor dynamics model.

I would like to acknowledge the financial support of the *Onderzoeksraad* of the University of Ghent, and the support of the Belgian National Science Foundation, in particular also M. Vanwormhoudt. I. Schepers and B. Willems provided technical assistance and D. Moelants helped in preparing figures for the final completion of the manuscript. He also assisted me with the evaluation of the TCAD model (Chap. 10). S. Slembrouck checked the language.

The book is dedicated to my friend, humanist, musicologist, and teacher J.L. Broeckx. His work on music aesthetics, in particular his book *Muziek, Ratio en Affect* (Metropolis, Antwerpen, 1991) has been a source of inspiration for my work.

My last words of thank go to Magda and Batist. Without their warmth and distraction, I would never have been able to explore this hitherto unknown world of musical imagery.

Ghent, March 1995 Marc Leman

Table of Contents

1. Introduction

This book is about schema theory, about how memory structures self-organize and how they use contextual information to guide perception. The schema concept has origins in philosophy (E. Kant), neurology (H. Head) and psychology (F.C. Bartlett, U. Neisser, J. Piaget) and is now generally accepted as a fundamental cornerstone in AI (Artificial Intelligence), cognitive science, and brain research [1.6].

Cognitive psychologists have come up with a paradigm for research about schemata in which music has been found to be an important domain of application. The paradigm, known as *cognitive structuralism* [1.7], is based on an analysis of similarity judgments between distinct objects. These judgments, processed with multi-dimensional scaling and hierarchical clustering techniques, suggest memory structures of perceptual knowledge. The mental maps – as schemata are alternatively called – are conceived as analogical structures of a *second order* isomorphism. That is, a structure in which the relations between the represented objects reflect the relations between the perceived real-world objects [1.8]. A structure for *first order* isomorphism would imply that the represented objects reflect the real-world objects instead of the relations.

The multi-dimensional structures for pitch and timbre [1.1–5, 9, 10] have been mapped out with results that have contributed to a better understanding of music perception.

The paradigm is relatively successful but has nevertheless a profound limitation which was the starting point of the present research. The problem can be summarized as follows: cognitive structuralism provides a method for the registration of the surface level of schemata, but it does not take into account the underlying dynamics of *emergence* and *functionality*. The organization of a control structure indeed tells little, if anything, about the underlying processing and functioning. How does a schema come into existence? How does it function in a particular perception task? The representational paradigm is static and insufficient for an explanation of the dynamics of sensorial and perceptive processes. The so-called "semantic roles" of musical objects are ignored or referred to in vague terms. It is indeed difficult – if not impossible – to represent them as fixed structural representations.

The aim of this book is to provide a foundation for the *emergence* and *functionality* of schemata by means of a case study in tone center perception. The methodological and epistemological foundations of this psychomorphological theory relies on an attempt to combine physiological acoustics (psychoacoustics) with self-organization theory (Gestalt theory). The schema concept, with its foundations in psychology and physiology, plays a central role in this.

2. Tone Semantics

This chapter gives an introduction to the problem of tone semantics, from a context-sensitive semantic point of view, as well as from a historical perspective. The chapter ends with an overview of recent achievements.

2.1 The Problem of Tone Semantics

While music semantics is concerned with musical meaning in general, *tone semantics* refers to the more specific problem of how tones, in a context of other tones, relate to each other and "behave" in a quite orderly and meaningful way. Tone semantics points to the way in which the human mind identifies tones and assigns them a functional relationship (such as leading tone, tonic ...).

Since perceptual meaning formation (apart from the cultural background and individual experience of the beholder) seems to be determined both by

– properties of the percept *and*
– properties of the context in which the percept occurs,

it is useful to distinguish between the *proper* meaning and the *context-sensitive* meaning of an object. The proper meaning – also *denotational* or *standard* meaning – of the word "chair", for example, points to a piece of furniture to sit on. But there are other possibilities. "Chair" may mean something to sit on, or to hold (as chairman), or just to look at (in the arts). The interpretation depends on the context: the sentence, the "discourse", and possible other – non-linguistic – events ("speech acts") that are associated within spatial and temporal constraints of the utterance.

The *context*, which embeds the object in an environment of other objects, may suggest meanings that differ from the common meaning. In speech, different contexts may lead to different interpretations of a word. Contexts often create or solve ambiguities, and the context-sensitive meaning is therefore an important aspect of meaning in natural language.

In music perception, the distinction between *proper* meaning and *context-sensitive* meaning is particularly relevant in that the so-called *direct meaning* formation is almost exclusively concerned with context. This applies especially to one of the most important building blocks of Western classical music: the tone. Tones in isolation have no proper meaning. They do not refer to something else. A tone points to nothing but itself and hence it is *self-referential*. The meaning of the tone, what makes a tone a *leading tone* or a *tonic*, is determined by a tone-context – a spatial and temporal environment of other tones.[1]

In Sect. 12.5 it is argued that tone semantics implies *automatic* and *direct* meaning formation – without explicit conscious processing. In that sense, the context-sensitive semantics in music has particular properties implying an interaction between three elements: (i) the object, (ii) the context in which the object appears, and (iii) the schema or structure that controls its perception. In this perspective one may wonder how both

- the meaning of a tone (or chord) can be determined by its context, while
- the context itself is determined by the constituent tones (chords).

In other words, what is determined by the context is itself part of that context, and as such, also contributes to the emergence of that context. An important question concerns the role of the schema and how it interferes.

The above interaction of object/context/schema sounds rather paradoxical from a static point of view and can therefore only be understood from a dynamic point of view. This is the core of a psycho-morphological theory of music. In this book, a provisional answer is given to the problem of the dynamics of context-sensitive semantics. The proposed theory is based on a case study in tone center perception.

2.2 Historical Background

The theory of tone semantics is associated with a cultural development. In Western Europe, a particular system of context-sensitive semantics has evolved over the centuries towards a more or less stable system, known as the tonal system. Theories about this system have evolved as well, and in this section a historical overview is given of the most important achievements.

In the sixth century BC, Pythagoras (ca. 500 BC) discovered that tone intervals could be represented by simple ratios between the lengths of strings. The interval of an octave was found to correspond with the ratio 2:1, that of a fifth with 3:2, and so on. By taking a string of a certain length, say S, one could prove that the tone produced by a string of length $S/2$ sounds twice as high. Similarly, a string with length $2S/3$ would produce a tone that

[1] *Wishart* [2.31] has noted that there are musical building blocks which operate at a low perceptual level and which *do* refer to objects outside music. Such percepts are often used as reference points in sound-transformation (e.g., interpolation). Of course, the musical context will partly determine the perception of these cues as well.

sounds one fifth higher, and so on. These ratios, which are now conceived of in terms of frequency ratios, were thought to express the relationships between celestial bodies.[2]

For centuries, music theory has been influenced by the Pythagorean fascination with numbers. This probably explains why mathematicians have been intensively involved with music theory. A most famous example is L. Euler (1707–1783), who tried to establish an arithmetic foundation for tone semantics following G. Leibniz's (1646–1716) idea of the "secret calculation of the soul". His "gradus suavitatis" (or degree of melodiousness)[3] can be considered the first step towards a computational theory of tone semantics [2.16]. Euler suggested that the degree of melodiousness depends on calculations made by the mind: the fewer the calculations, the more pleasant the experience. A low number of calculations leads to a high value for melodiousness, while a high number of calculations yields a low value.

This principle is implemented by a numerical technique based on the decomposition of natural numbers into a product of powers of different primes. If $p_1, \ldots, p_n$ are different primes and $e_1, \ldots, e_n$ are different powers, then any natural number a can be expressed as

$$a = p_1^{e_1} \cdots p_n^{e_n} .$$

The degree of melodiousness is expressed by

$$\Gamma(a) = 1 + \sum_{k=1}^{n} e_k(p_k - 1) \tag{2.1}$$

with

$$\Gamma\left(\frac{a}{b}\right) = \Gamma(a * b)$$

The latter equation is introduced to deal with the rational numbers expressing intervals. For example, the degree of melodiousness of the fifth is

$$\Gamma\left(\tfrac{3}{2}\right) = \Gamma(6) = 4.$$

The function produces the values of the intervals given in Fig. 2.1. The associated table contains three columns: the intervals in ratios (prime, minor and major second, minor and major third, fourth, tritone, fifth, minor and major sixth, minor and major seventh), the "gradus suavitatis" (or $\Gamma(interval)$), and its normalized inverse (which is plotted). The plot is called a *tone profile.* By shifting the patterns over all tones of the scale it is easy to see that there are 12 such profiles, one for each reference tone on the scale.

Nowadays, the principle of economy of thought ("Occam's razor") which underlies Euler's model is no longer accepted as a foundation for perception.

---

2 Although metaphysics does not perturb the scientific mind any longer, some authors argue that even recent developments in the quantification of empirical reality should be considered achievements of Pythagorean tradition. "We are all Pythagoreans" says *Xenakis* [Ref. 2.32, p. 40].

3 In his *Tentamen novae theoriae musicae* (1739).

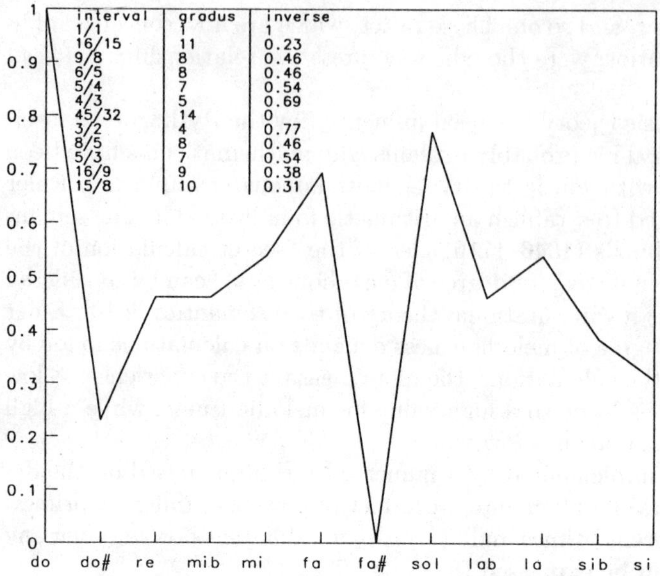

Fig. 2.1. Tone profile based on the "gradus suavitatis" (Euler). The intervals are given, together with the gradus (calculated according to (2.1)). The inverse (plotted) is scaled with respect to column 1

Numerical recipes or purely arithmetic models do not incorporate the psychophysical foundations of pitch perception and hence cannot serve as a basis for a theory of tone semantics. For instance, the phenomenon of the enlarged octaves provides a counter example to Euler's model.

The ratios of enlarged octaves are complicated because they deviate from the pure ratio. Yet, they sound consonant ("pleasant") and they are used all over the world in all cultures. The phenomenon of the enlarged octave is now explained at a physiological level, by referring to the refractory delay in peripheral neurons and/or to enlarged spacing of harmonic excitation patterns [2.5]. The deviation from the pure ratio is in contradiction with Euler's model because stretched octaves involve too much computation. An additional point of criticism is that this model – which still serves as a paradigm for arithmetic accounts of tone relations – does not take into account any notion of *context*. The latter is perhaps most important for a theory of tone semantics.

The first description of a context-sensitive semantics was the work of a musician. Musical practice is indeed grounded in context-dependent activity. Knowledge of the context-sensitive character of tones ("a tone sounds good in one context and bad in another") is something to be learned by experience – more an art than a science. The most influential theory has been formulated by J.-P. Rameau (1683–1764) in his *Traité de l'Harmonie* [2.18]. This treatise comprises a theory of harmonic intervals and chords, and a theory of chord progressions. The former is based on acoustical principles worked out by G. Zarlino (1517–1590). The latter comprises perhaps the most original contribution.

According to Rameau, the intervals of octave, fifth, major third and minor seventh are fundamental because they appear in the series of natural overtones. Chords, in which these harmonic intervals are used, are assumed to support a *fundamental bass* ("Basse Fondamentale"). This bass is not a real audible low tone, but a construct of Pythagorean origin to explain the harmonic function of the interval. Accordingly, the behavior of tones and chords in a tone context is described through the interaction of three features [2.2]:

– the acoustical properties of tones,
– the construct of the "Basse Fondamentale", and
– the resolution of dissonants.

In the terminology of computer science, the model can be characterized as a *symbol-based constraint-satisfaction system.* Symbol-based expert systems for harmonization or analysis [2.15] still rely on this important work.

Rameau's model provides interesting recipes for musical practice, but it has severe limitations as well. His description of context-sensitive semantics is phenomenological and does not take into account the underlying sensorial and perceptive foundations. Its ultimate foundation is art and intuition – not science and experiment. Therefore, it is not a good explanatory model for tone semantics.

The next important achievement in tone semantics theory was the introduction of psychophysics and physiological acoustics. It was introduced by H. Helmholtz (1821–1894) in *Die Lehre von den Tonempfindungen* [2.6]. Tone semantics is interpreted in terms of a similarity, based on the principle of common partials: chords or tones that have common partials sound less rough (dissonant) than those that do not have common partials. This phenomenon is explained as the result of interferences of waves in the ear.

The roughness of intervals is shown in Fig. 2.2a,b. Figure 2.2a displays a curve computed by Helmholtz. The horizontal axis compares the intervals which are less than one octave. The vertical axis gives the value for roughness (dissonance). Dissonance gets a high value, consonance a low value. Figure 2.2b has been deduced from Fig. 2.2a by taking the intervals identical to Euler's (first column). The data for dissonance (second column) have been taken from a table that corresponds with Fig. 2.2a, rescaled (third column) and plotted. As Fig. 2.2a suggests, slightly other values are obtained with the equal temperament scale instead of pure ratio. But the general tendency remains the same.

It is important to remark that the tone quality of instruments (the amount and energy of the harmonics) has an important effect on the tone profile. Instruments with many strong harmonics produce more dissonance than instruments with few and weak harmonics. Helmholtz computed Fig. 2.2a by taking a tone quality model of a violin. Nevertheless, the dissimilarity be-

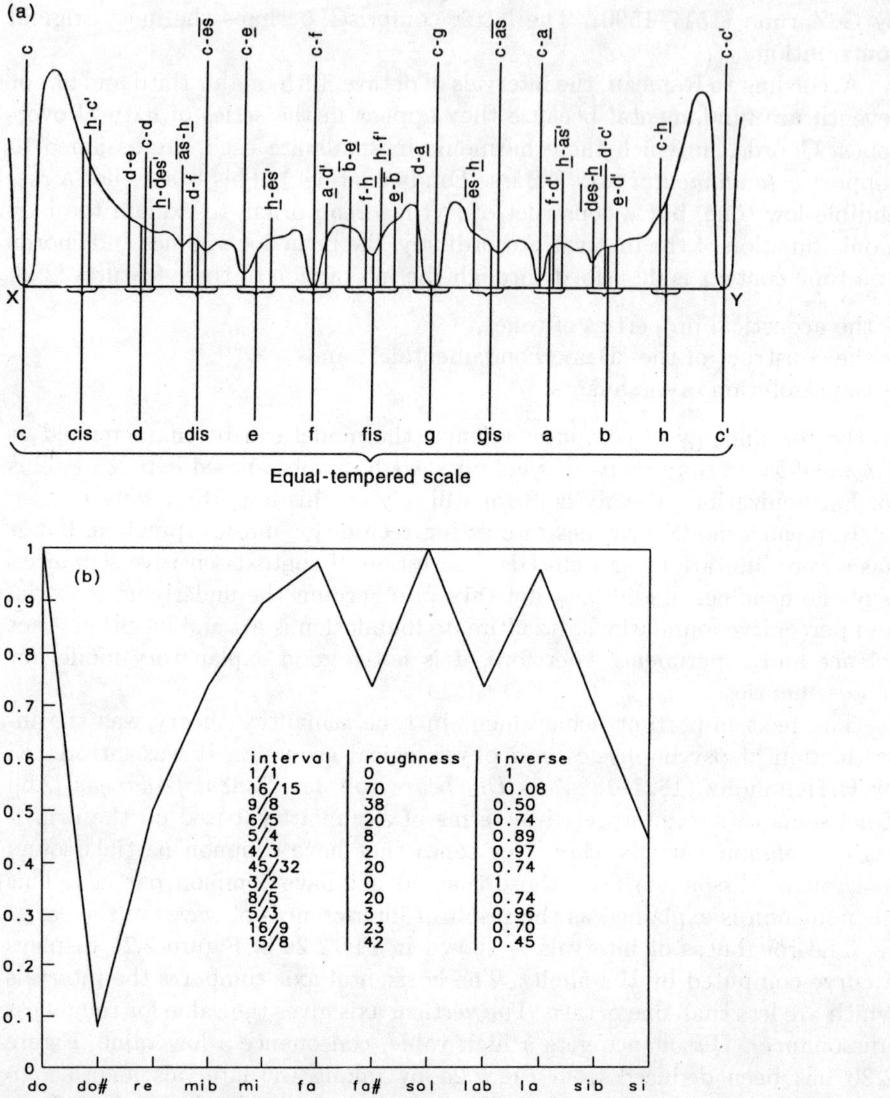

Fig. 2.2. Tone profiles based on roughness: (a) table calculated by Helmholtz [2.6], (b) the curve reduced to the intervals used by Euler (Fig. 2.1). The plotted curve is the inverse of roughness, scaled to 1

tween the tone profile obtained by Euler (Fig. 2.1), and the one obtained by Helmholtz (Fig. 2.2a) is remarkable.

From a musical point of view, it can be argued that the psychophysical approach, based on similarity relationships between tones and chords, provides no firm basis for the explanation of context-sensitive meaning. Musical contexts indeed involve learning processes which introduce a cultural factor, whereas the definition proposed by Helmholtz reduces this factor to sensorial properties.

Many music researchers have since realized that sensorial principles alone are not sufficient to explain tone semantics. Helmholtz was therefore criticized for having restricted the phenomenon to purely sensorial aspects. Although much of the criticism was probably misplaced [2.27, 29] – Helmholtz was indeed aware of the restrictions of his theory –, the music theory of the Post-Helmholtz area evolved more and more away from a psychophysical approach towards a phenomenological approach.

C. Stumpf (1848–1936) asked why we hear one single pitch when a key is pressed on the piano, instead of the different partial tones that are analysed by our ear [2.23, 24]. He argued that *tone fusion* played a central role in pitch perception. Tone fusion was considered a cognitive principle, while the frequency analysis was a sensorial principle. His *Tonpsychology*, written in the spirit of his teacher F. Brentano (1838–1917), announces already a Gestalt theoretic approach to perception, one in which the tone is considered the Gestalt of a tone complex. His experiments in tone fusion result in a diagram or tone profile of "Verschmelzungsstufen" (degrees of fusion) (Fig. 2.3). The diagram was obtained by asking listeners how many tones are heard while playing an interval. The intervals are marked by ratios.

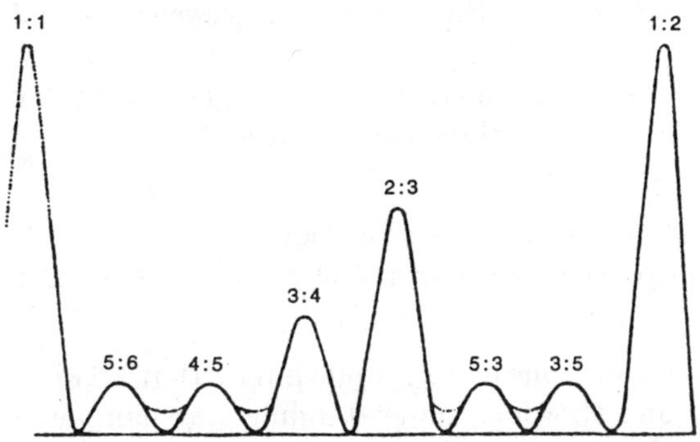

Fig. 2.3. Tone profile based on psychological data (Stumpf)

Another representative of the Post-Helmholtz tradition in systematic musicology is E. Kurth (1886–1946). He explains tone semantics by concepts such as *force* and *energy.* According to *Kurth* [2.13], a tone has nothing musical by itself, but gets a musical meaning in a musical context. The vitalistic metaphysics in which this idea is expressed sounds a bit archaic today, but it is important as a metaphor for a system theoretic account of music perception. The claim that the music makes the tone ("Musik macht den Ton") reflects the idea that the semantic properties of tones are dominated by the context in which they appear.

From then on, tone semantics was strongly associated with Gestalt theory. As A. Wellek (1904–1972) noted, the meaning and nature of a tone depends on the construction of a musical culture, and in this sense also on a given point of human development and history. Experiencing a tone in a musical sense means to put the whole music into it [Ref. 2.30, p. 81].

Many attempts of the systematic musicologists, however, remained programmatic or metaphorical and could not, despite to the brilliant work of Helmholtz, provide an operational account. Stumpf, for example, posited "specifische Synergien" as physiological foundations of tone fusion [Ref. 2.24, p. 214] but failed to provide a model. Other researchers even strongly opposed the empirical approach. H. Riemann (1849–1919), for example, argued in favor of a *music logic* and disregarded the inductive methods of tone physiology and tone psychology. His attempts to ground tone semantics in a logic of tone imagination resulted in a dualistic (major/minor) system which, according to *Dahlhaus* [2.2], is incoherent and dogmatic.

To summarize: this short historical overview illustrates different approaches to tone semantics. Consonance theory was associated with mathematics and adopted a quantitative approach. Musical practice discovered the role of context but provided descriptive models rather than explanatory models. With the achievements of Helmholtz and Stumpf, music theory underwent a radical change. It became obvious that a solution to the problem had to come from the empirical sciences: physiology, psychoacoustics, and the psychology of music cognition. The new approach exchanged intuition by rigorous scientific methods and made a clear separation between art and science.

At the end of the nineteenth century it was realized that a genuine theory of tone semantics had to take into account three constraint levels:

1. sound acoustics,
2. processes of sensorial perception and Gestalt formation,
3. the cultural environment (in particular the distribution of tones in time frames).

Tone semantics was seen as the outcome of a dynamic interaction between these levels but its description and thorough understanding was hampered by a lack of operational models. In the first half of this century, this impediment gave systematic musicology its programmatic character. The recent developments in psychology, neurobiology, and computer science, however, provide powerful tools and concepts by which complex dynamic systems become testable. It is therefore important to review two recent achievements in psychology and discuss their relevance in the framework of tone semantics.

2.3 Consonance Theory

Consonance is traditionally considered an essential feature of tone semantics. According to Helmholtz, consonance is the undisturbed flow of simultaneous tones occurring when certain tones form perfectly definite intervals (such as the octave or fifth). When these intervals do not occur, beats arise [Ref. 2.6, p. 204]:

> that is, the whole compound tones, or individual partial and combinational tones contained in them or resulting from them, alternately reinforce and enfeeble each other. The tones then do not coexist undisturbed in the ear. They mutually check each other's uniform form. This process is called *dissonance.*

When beats follow each other faster and faster, they fall into a peculiar pattern of dissonance called *roughness.*

Helmholtz's statement that the sensation of roughness results from the interference of waves has been confirmed in recent studies. It was found, however, that the frequency resolution of the ear is somehow constrained: only tones that fall within well-defined frequency groups (called critical bands) interfere [2.34]. Tones that fall outside these areas do not interfere, and hence do not cause the sensation of roughness. Other effects, such as beats and masking (the suppression of one frequency by another) also occur in these zones.

Zwicker and *Fastl* [2.33] assume a constant bandwidth of 100 Hz for frequencies up to 500 Hz and a relative bandwidth of 20 % for frequencies above 500 Hz. But depending on the method used, the results are somewhat different. Recent estimates give smaller bandwidth for frequencies below 500 Hz [2.17]. For musical purposes, however, the width of the sensitive bandwidth (or zone) is taken to be about 1/3 octave (minor third). Figure 2.4 shows some estimates for critical zones. The dotted line shows the "classical" curve, while the full line shows the more recent estimates. Roughness falls within the critical bandwidth.

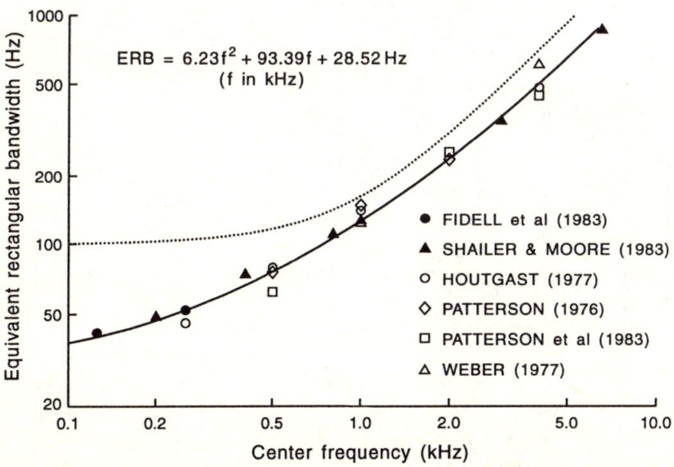

Fig. 2.4. Estimations of critical frequency zones [2.17] (By permission of the authors and publisher)

With the arrival of electro-acoustics, Helmholtz's theory could be verified much more accurately. Figure 2.5 shows results of a study by *Kameoka* and *Kuriyagawa* [2.7] in which the degree of dissonance between two tones with varying harmonics was measured experimentally. The upper curve, based on simple tones, displays the dissonance of tone intervals that vary over one octave. The second curve results from tones with a first and second harmonic, and so on. The lower curve compares the dissonance of two complex tones, each with 6 harmonics added. The figure shows that intervals composed of tones with many harmonics sound rougher (dissonant) than tone pairs in which tones have few harmonics. This effect is explained by the interference of the harmonics within the critical zones. Notice the similarity with Helmholtz' calculations shown in Fig. 2.2a,b.

One should be careful about a musical interpretation of these results. The experiments are carried out with isolated tones and any reference to a musical context is explicitly avoided. The figures are therefore limited to a picture of the *sensorial* aspects of consonance. The history of music, on the other hand, suggests that sensitivity to consonance and dissonance is *learned.* What was experienced as a dissonant interval in the 16th century was later considered to be consonant. Therefore, it is difficult to maintain that the concept of consonance has a unique sensorial basis and that a distinction

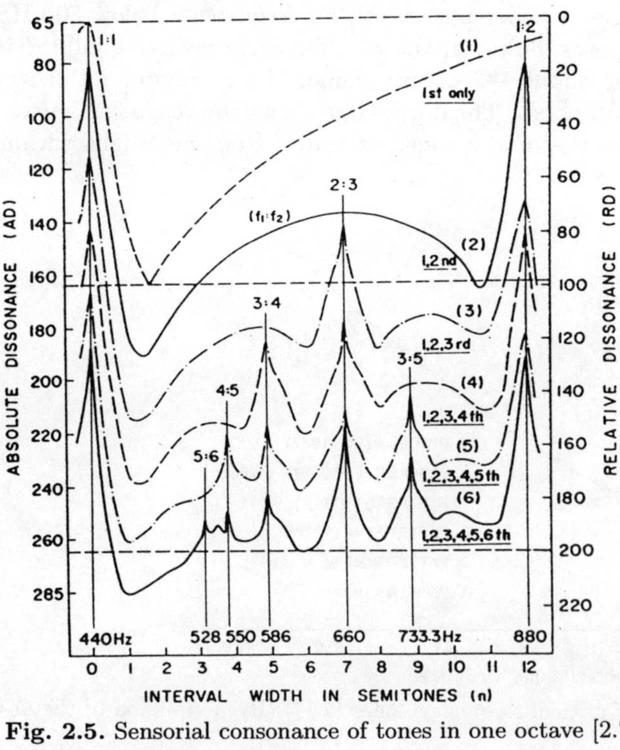

Fig. 2.5. Sensorial consonance of tones in one octave [2.7]

should be made between *sensory* consonance (the inverse of roughness) and *musical* consonance [2.27]. Expressions of the latter are found in the identity of chroma (toneness) between tones in a sequence, in the similarity between inverted chords, or in the nature of some harmonic progressions.

In general, musical consonance is best considered a derivative of tone semantics. Indeed, the most important difference between sensory consonance and musical consonance has to do with context. Musical consonance is the resultant force of sensory consonance embedded in a musical context. Actually, Helmholtz was quite aware of the role of context. He says [Ref. 2.6, p. 229]:

> if the boundary between consonance and dissonance has really changed with a change of tonal system, it is manifest that the reason for assigning this boundary does not depend on the intervals and their individual musical effect, but on the whole construction of the tonal system.

He did not, however, formulate a compound theory of pitch perception in which context is taken into consideration. As mentioned above, this has been a source of great confusion and discussion.

Sensorial consonance is an important aspect of pitch perception, and must be part of any model of tone semantics. The present model takes sensory consonance into consideration by means of an auditory model (Chap. 5).

2.4 Cognitive Structuralism

Studies of context-dependent pitch perception were carried out in the late seventies by *Krumhansl* and *Shepard* [2.11]. C. Krumhansl has given a good overview in her book on the cognitive foundations of musical pitch [2.9].

In accord with the cognitive structuralist point of view, the aim was to quantify the structure that underlies the functional relationships between tones in a particular piece. The listeners were asked to judge, on a scale from "very similar" to "very dissimilar", the similarity between a pitch and a given tone context. For example: a cadence in F-major is played and the listeners rate the similarity between this given tone context and a pitch that follows it. The experiment is repeated for all tones of the chromatic scale and the judgements are stored. The experiment can be extended to different contexts and different test objects (chords instead of tones).

The responses of the subjects are represented as tone profiles. When tone profiles are analysed with a mathematical technique known as multidimensional scaling [2.12] it becomes possible to give a graphical representation of the mental representation that underlies the particular task.

Before summarizing the main results of this experiment it is perhaps useful to mention that *Shepard-tones* have been used. These tones, the nature of which is discussed in Sect. 3.3 have special properties which allow a reduction of musical parameters. Shepard-tones have the property of being circular:

going up one octave results in a tone with the same pitch as the starting one. By using Shepard-tones, the influence of *height* on harmony is neutralized and the perceived frequency range is reduced to one octave. A chord and its inversion have exactly the same perceptual effect. Accordingly, the tone profiles span one octave.

Figure 2.6a,b depict the tone profile of the C-major context, and the tone profile of the C-minor context. It is important that the reader has a good understanding of these pictures, since they provide a basic reference for later discussion. The numbers refer to the mean ratings on a scale from 0 to 7 (dissimilar-similar). One obtains the profiles for all the other contexts by shifting the pattern of Fig. 2.6a one unit to the right. The unit which goes out of the diagram at the right is wrapped back on the left side. Starting with C-major, one thus obtains the tone profile for C♯-major, D-major, and so on. A similar operation can be carried out on the pattern of C-minor (Fig. 2.6b). There are 24 different patterns that can be obtained through rotation: 12 for the major context and 12 for the minor context.

A multi-dimensional scaling analysis of these 24 patterns leads to the structure depicted in Fig. 2.7. The structure is a torus, which means that the upper and lower sides connect, as well as the right and left sides. Each label points to the tone center of the corresponding context.[4]

One observes that major and minor tone centers are related to each other in circles of fifths: **C, G, D, A, E, B, ...** and **c, g, d, a, ...** In addition, each major is flanked by its parallel minor and relative minor. For the tone center of **C** this is **c** and **a**, respectively.

Two important structural principles of the mental representation are:

- the structure is *analogical* in the sense that relations between represented objects reflect the relations between perceived objects,
- the structure is *topological* in that the similarity relationship is translated into distance: short distance stands for *similar*, long distance stands for *dissimilar*. Related tone centers (e.g., **C** and **G**) appear close to each other, while those that are unrelated (e.g., **C** and **F♯**) appear distant from each other.

There is an alternative way to represent the data of Fig. 2.6a,b, as is shown in Fig. 2.8a,b. Figure 2.8a displays the similarity of all contexts with

[4] Concepts such as *tone profile*, *tone center*, *tonality* and *key* denote different things, but their meaning is related. They should therefore be used with care. The tone contexts used in Krumhansl's application evoke the sense of a *tone center*. Strictly speaking, a tone center is not a synonym for tonality or key. A tone center is a psychological category while a tonality or key is a music theoretical construct – often associated with a scale. A *tone center* refers to a stable perception point and can be generated by a tone sequence that stands for a key or tonality. This is typically a cadence. The notion of *tone context* is more general. In the experimental setup, the tone context generates a strong reference to a tone center, but this is not necessarily so. In music, a tone context is often ambiguous. Cadences are used to make contexts less ambiguous.

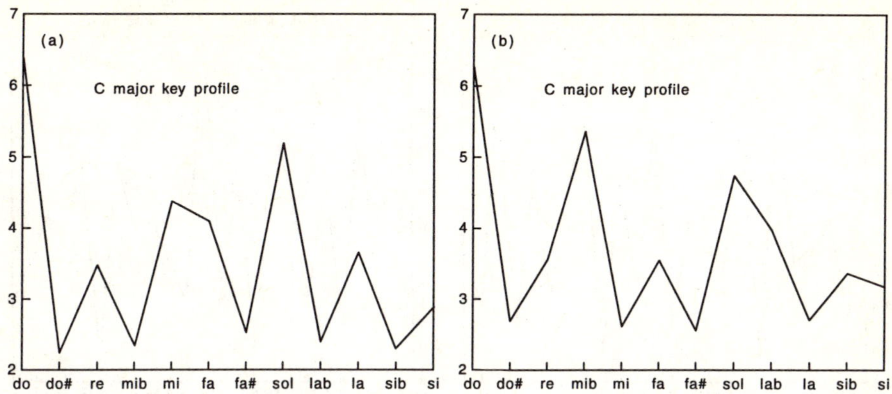

Fig. 2.6. Tone profiles of (a) the C-major and (b) C-minor key (Based on [2.9])

respect to the context of **C**. Figure 2.8b does the same with respect to **c**. The tone centers that are most similar to **C** are **F, G, c, e** and **a**. These are the centers that are closest in distance to **C** in Fig. 2.7. These figures characterize the underlying similarity structure of the mental representation. The organization is similar to tonality structures known in music theory [2.14, 21]. Is it possible to show that the structure emerges from just listening to music? How is it used in a tone center perception task?

Apart from experimental studies in perception, which give an insight into the structural aspects of mental representations, other results have been based on the analysis of musical performance. The aim here is to quantify the rules that underlay musical performance, in particular the transition from a score to a musically acceptable output. Studies in musical performance indeed show

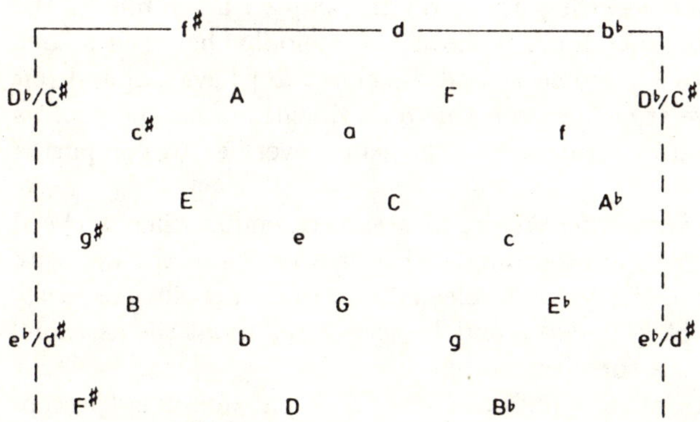

Fig. 2.7. Multi-dimensional scaling solution of 24 tone profiles [2.9] (Copyright 1982 by the American Psychological Association. Reprinted by permission of the publisher)

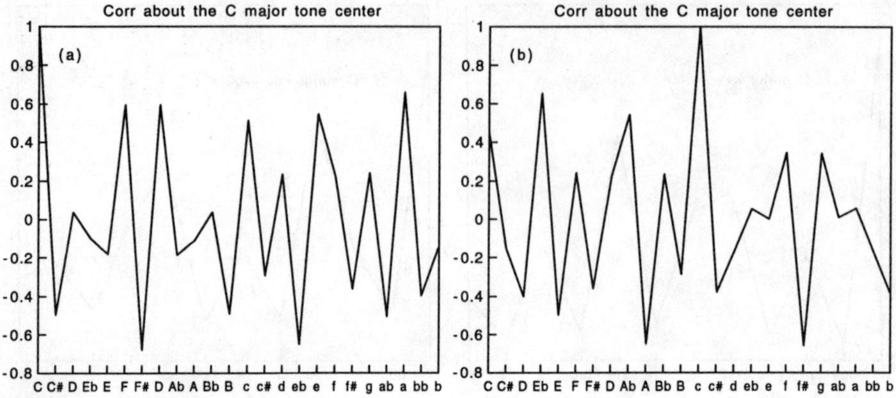

Fig. 2.8. Correlations among tone profiles: (a) all tone profiles are correlated with the C-major tone profile, (b) all tone profiles are correlated with the C-minor tone profile. (Based on [2.9])

that musical expression relates to accents in tuning, duration, and loudness. These parameters are constrained by context [2.1, 3, 4, 19, 20, 28].

The paradigm adopted is based on an *analysis-by-synthesis* method. Listeners, confronted with computer generated performances of a score, are asked to suggest improvements to the computer model. The suggestions are translated into a parameter set of performance rules and fine-tuned by repeating the procedure until acceptable results are obtained.

Sundberg et al. [2.25] gives a good overview of research in this area. One important category of rules pertains to the prominence of tones in the harmonic environment. This prominence is quantified by the notion of *melodic charge*. Increments in duration, amplitude and vibrato, and change in intonation are added to a tone according to its relationship with the root of the prevailing chord. The profile of Fig. 2.9 shows the melodic charge in a tone context which emphasizes **C**. *Sundberg* and *Fryden* [2.26] have pointed out that there is a high correlation between this quantification and the profiles of the major contexts used by Krumhansl. This can be verified by comparing Fig. 2.6a with Fig. 2.9.

Another rule, called *harmonic charge*, creates crescendos when a chord of higher harmonic charge is approaching and decrescendos in the opposite case. Sundberg and Fryden show that a relevant correlation coefficient exists with the research results of *Krumhansl* and *Kessler* [2.10] about the listeners' ratings of major chords in a tonality context.

The fact that similar data are obtained from different domains (perception and performance) and by different methods suggests that the notion of context-sensitive semantics is an obvious one and it is consistent.

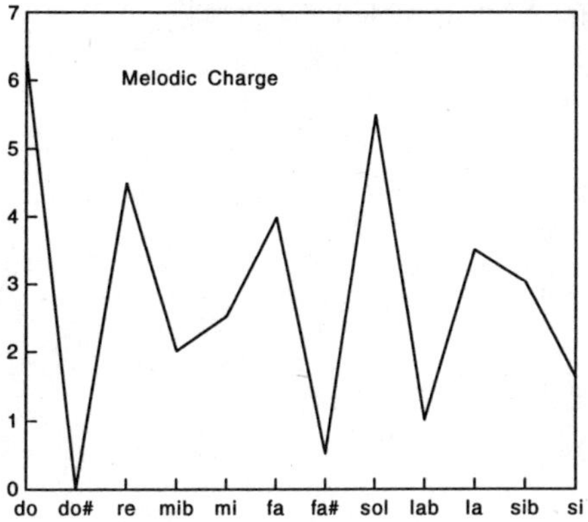

Fig. 2.9. Tone profile based on "melodic charge" (Based on [2.25])

2.5 The Static vs. Dynamic Approach

The above achievements of consonance theory and cognitive structuralism have contributed to a better understanding of the organizational aspects of the schema that underlies tone semantics. But as suggested in Chap. 1, this is not sufficient as long as there is no proper explanation for the schema in terms of its emergence and functioning control structure.

The failure of a purely static approach is perhaps best illustrated by Shepard's influential paper on structural representations of musical pitch. Shepard's original idea was that [Ref. 2.22, p. 350]:

> in order to preserve the musically important structural relations between tones (the musical intervals of the octave, fifth, and so on) and in order to preserve invariance under the musically important transformations (e.g., transposition), the cognitive representation of musical pitch must have properties of great regularity, symmetry, and transformational invariance.

This idea led to the development of ingenious structures for pitch representation, of which an example is shown in Fig. 2.10. The double helix model is the result of careful deduction based on basic properties of pitches. Its empirical correlate, on the other hand, refers to the quantified schema of Fig. 2.6a,b. But it turned out that the deductive approach was not without problems, as Shepard observes [Ref. 2.22, p. 385]:

> Still remaining to be fully elucidated are the rules of projection that map such an abstract, symmetrical structure into the more concrete and perhaps asymmetrical structure representing the perceived relations between tones within the context of a particular musical key-between, that is, the unique tonal functions (tonic, dominant, subdominant, leading tone, etc.) relative to that key.

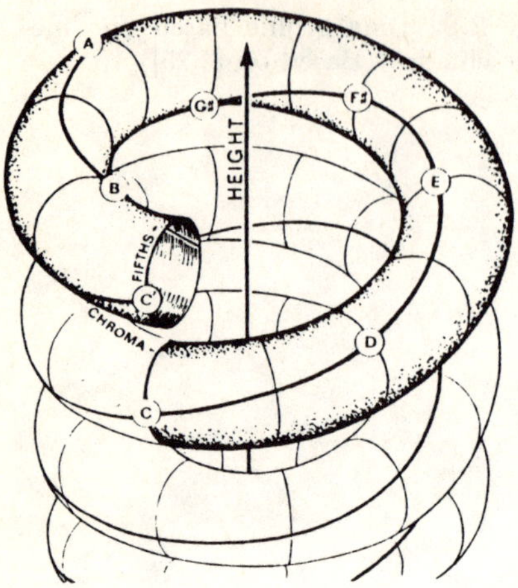

Fig. 2.10. Double helix [2.22]

In other words, it is fully acknowledged that the different semantic roles of tones may be difficult to represent in structural representations. Such representations indeed have a static character and it is difficult to conceive of ways in which the dynamics of tone semantics can be incorporated within these structures. It is therefore better to consider

- a carrier for the representational structures, and
- a dynamic system within which these structures emerge and function.

In the cognitive structuralist approach, both the carrier and the dynamics are not interpreted. This separation of representation and system is one of the weak points of the approach. It has furthermore led to the hypothesis that listeners have abstracted and internalized the patterns of tonal distributions [Ref. 2.8, p. 24, Ref. 2.9, p. 76]. This hypothesis, based on the observation that the tone profiles have a higher correlation with the statistics of tone distributions in tonal compositions than with the roughness profiles discussed in Sect. 2.3, is underdetermined.

The cognitive structures for tone perception are supposed to be learned, but the statement that listeners internalize the summary statistics of tones in music is far too general. Such structures can only be fully understood if we understand something about their development. Hence, learning mechanisms should be studied, as well as the musical conditions under which schemata emerge.

2.6 Conclusion

For centuries, people have been fascinated by the relationships between tones. Contributions to a better understanding of these principles have been made both by scientists and musicians. Of particular relevance have been the treatises on musical practice and the shift towards psychophysical explanations.

On the other hand, it is only recently that contextual factors have been studied in a scientific way. Former studies in dissonance perception precisely avoided the influence of context and concentrated on isolated factors. Cognitive structuralism aims to quantify the context-sensitive semantics of pitch perception into mental representations. But both the carrier for such structures, as well as the underlying control and learning principles, have not been considered. The distinction between representation and carrier system is a weak point that can be overcome with advanced computational tools. In the following chapters of this book, it is shown that such an approach is indeed possible.

3. Pitch as an Emerging Percept

This chapter is about the decline of the traditional phenomenological approach to pitch perception. Auditory experiments with ambiguous stimuli provide arguments in favour of a model in which pitch is seen as an emerging outcome from auditory information processing, rather than in terms of phenomenological attributes.

3.1 The Two-Component Theory of Révész

Music theory relies on a pitch concept that is often associated with the two-component theory of *Révész* [3.13]. In this theory, pitch is seen as a concept with two independent attributes:

- the first varies with frequency and shows itself in the phenomenon of *height* (a pitch is going up or down). The attribute has the nature of being *linear* or *monotonic*.
- the second attribute is repeated every octave and is called *toneness*.[1] Toneness points to the position within the octave and is *circular* or *cyclic*: it repeats each octave and – since there are 12 tones in one octave of the equal tempered scale – has a modulo-12 function.

This distinction is based on two pillars: the phenomenon of octave-equivalence, on the one hand, and research in hearing on the other.

- The phenomenon of the octave equivalence was considered to be the "alpha and omega" of the musical system and the hearing system [Ref. 3.13, p. 71]. Its foundation was thought to be reflected in the repetition of the note names at every octave.

[1] Other terms for toneness are: tone chroma, tone quality, tonality, and pitch class. The latter term, however, has a foundation in music analysis based on set theory [3.7] and is probably less useful here. *Tonality* deals with the function of tones in a musical context and is also less appropriate, because we are dealing here with the phenomenal structure of tones in isolation. *Tone quality* is sometimes used to denote the timbral character of tones, while *tone chroma* suggests that the phenomenal structure has absolute properties, like colors have. Unlike colors, however, the mutual relations are isotropic rather than absolute. The term "tonalness" is often used in contrast to "noisiness". The tonalness of a tone is higher when there are more components of the harmonics available [3.20].

- In addition, research in musical hearing has shown that persons with an absolute hearing capacity sometimes make mistakes in the identification of the octave register but never in the names of the notes. It was concluded that such subjects know the pitch on the basis of toneness while persons with a relative hearing capacity determine pitch basically on the basis of its height.

The two-component theory provided a foundation to music theoretical findings that were circulating in the second half of the nineteenth century. Figure 3.1 shows a graphical representation of the relationship between the attributes. In fact, this representation is nothing less than an attempt to capture the mental representation of pitch. The monotone component corresponds with the vertical axis, while the circular component corresponds with the horizontal plane. One rotation of a point on the helix corresponds with one octave up or down, depending on the direction of rotation. Unlike the mental representations of the cognitive structuralist paradigm, Fig. 3.1 is not deduced from a mathematical analysis of empirical data. The original helix-structure is an artifact based on intuition and musical practice but recent research [3.12, 16] reveals that a helix-like structure can be deduced from a multi-dimensional scaling of empirical data.

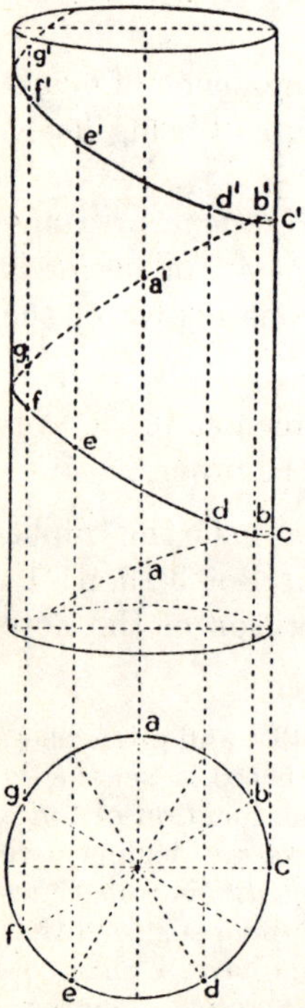

Fig. 3.1. The helix as idealized mental representation [3.13]

3.2 Attribute Theory Reconsidered

From the point of view of musical semantics, the relevance of the two-component theory is questionable – not only because it is about isolated tones in stead of tones-in-a-context. There are other fundamental reasons why this theory is too simple. First of all, the basis of the two-component theory – the principle of octave equivalence – is no longer a valid foundation of pitch perception. Psychoacoustical studies [3.1, 10, 12] show that circularity is more general and not necessarily restricted to the octave: it can be smaller or greater than the octave and much depends on the overtone structure of the tones. The results suggest that supremacy of the octave in Western music is culturally bound – probably inspired by the introduction of instruments with clear and rich overtones, such as violins and organs.

Moreover, toneness and height are not really orthogonal, as Fig. 3.1 suggests. The so-called illusions of perception (Sect. 3.5) suggest that toneness already incorporates aspects of height. Both aspects cannot be isolated strictly and are therefore not orthogonal. Besides toneness and height, work by *Nakajima* et al. [3.11, 12] suggests that there might even be a third phenomenological attribute based on temporal induction effects.

This shows that the concept of pitch is too complex to be represented by an attribute model. The *attribute theory* supports the notion of a tone as an abstract concept to which the attributes of height and toneness (but also: duration, timbre, loudness and dynamics) are assigned. Studies in psychoacoustics and music perception suggest however a more subtile truth: height, toneness and time are but *secondary* properties or *emerging* properties of perception. Their appearance is to be explained by more fundamental principles of auditory information processing.

Musicological research has felt itself quite comfortable with the attribute model until computational models revealed its vulnerability [3.8]. The focus on phenomenological attributes reduces the musical information to a skeleton of "notes-in-a-score". As a result, one looses the body of the musical signal: what makes music constrained, what gives music its semantics. In music theory, whose foundation is ultimately in perception, this cannot be justified since the skeleton provides insufficient information to explain inherent musical forces. The underlying assumptions of the attribute theory are therefore difficult to maintain in the light of recent developments.

3.3 The Shepard-Tone

Some convincing counter examples of the attribute theory have been based on Shepard-tones. A short description of the nature and perceptual effects of Shepard-tones is instructive here; the tones are used in studies of tonality perception (Sect. 2.4) and also in many chapters of this book.

The Shepard-tone is a signal that can be characterized by the following spectral properties:

- it contains sinusoidal tone components at intervals of one octave,
- the spectral energy of the components is weighted such as to enhance the region between 500 Hz and 1000 Hz.

The spectral envelope sometimes takes the form of a bell, as in Fig. 3.2, but the precise form is not very important. The perceptual effects of octave-component signals are rather robust.

It was R. Shepard who in 1964 built a tone which, since then, has been called the Shepard-tone. Shepard was inspired by the supposed orthogonality of height and toneness (Sect. 3.1) and wanted to study the effects of reducing the linear component in favour of the cyclic component. By projecting the helix (Fig. 3.1) onto a circle, it would be possible to hear pure toneness. The relationships between the pitches would then become relative in that there would be no highest or lowest tone but only tones in a relative position to each other. The signal was computed with one of the first digital sound compilers [3.9].

There are a number of ways for calculating Shepard-tones. This section gives the version adopted from *Shepard* [3.15]. The formula for calculating the frequencies is given by

$$F(t,c) \quad = \quad F_{\min} \; * \; 2^{[(c\,-\,1)\;*\;t_{\max}\;+\;t-1]/t_{\max}} . \tag{3.1}$$

The chromatic scale has 12 tones so that $t_{\max}$ is 12. $F(t,c)$ is the frequency of component c of tone t. $F_{\min}$ is the frequency of the lowest component of the first tone, such that $(F_{\min} \;=\; F(1,1))$.

The amplitudes of the components are calculated by

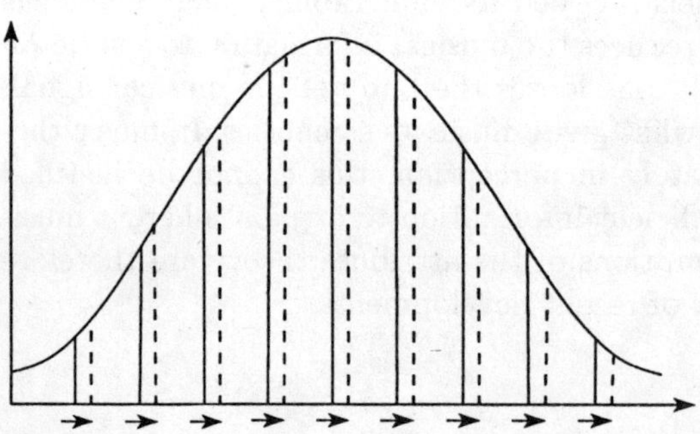

Fig. 3.2. The spectrum of the Shepard-tone is shown on a log2-frequency axis in function of the amplitude. The ten (full) vertical lines are the octave components. The dotted vertical lines represent a shift of the frequency components to a higher region. The amplitude in the higher frequency region becomes slightly lower, while it becomes higher in the lower frequency region. The bell-shaped envelop represents a filter in the spectral domain

$$L(t,c) = L_{\min} + (L_{\max} - L_{\min}) * \frac{1 - \cos\theta(t,c)}{2}, \qquad (3.2)$$

where $L(t,c)$ is the amplitude of component c of tone t, $L_{\min}$ is the minimum amplitude, $L_{\max}$ is the maximum amplitude.

The function $\theta(t,c)$ is defined by

$$\theta(t,c) = \frac{2\pi * [(c - 1) * t_{\max} + t - 1]}{(t_{\max} * c_{\max})}. \qquad (3.3)$$

The angle θ of the circle segment is defined by c and t. When $c = 1$ and $t = 1$, $L(t,c)$ reaches a maximum because $\cos\theta(t,c) = -1$.

It is known from psychoacoustical experiments that if components exceed the critical band, then the general loudness is approximated by the sum of the individual loudnesses [3.20]. Equations (3.1–2) guarantee a constant loudness for different Shepard-tones.

3.4 Paradoxes of Pitch Perception

Experiments with Shepard-tones have contributed to a decline of the phenomenological approach to pitch perception. The effect of Shepard-tones are indeed weird when considered from the phenomenological point of view. In the literature these effects are often called *illusions* and *paradoxes*, but they are perfectly comprehensible from the viewpoint of auditory information processing.

The main perceptual effect of a chromatic series of Shepard-tones is characterized by *isotropy*: each element on the circle has a neighbor in the clockwise direction that sounds higher and a neighbor in the counter clockwise direction that sounds lower. Playing a Shepard-tone one semitone higher is realized by shifting all octave components towards the higher frequencies (Fig. 3.1). In the equally tempered scale – which is always used in this book – the components will coincide with the original tone after 12 steps. Under conditions where the pitches are heard in isolation, both have equal height. Although this effect might appear musically counterintuitive, it is easily explained by the fact that the spectrum of the second tone is exactly the same as that of the first tone. Thus, by playing a chromatic scale of one octave, one ends with a tone that has the same height and toneness as the one started from. Other effects are also based on this principle.

Shepard-tones have been the subject of exhaustive studies by D. Deutsch and collaborators [3.2-6]. The following results have been obtained:

1. Isotropic height assignments are guided by the principle of *proximity*: Consider Fig. 3.3: a **RE** followed by a **RE**♯ (or **MI**♭) is up, a **LA** followed by a **SOL** is down, a **SOL** followed by a **MI** is down.
2. Proximity is switched off for tritone intervals: the perceived height direction of Shepard-tone pairs (up or down) varies systematically in function

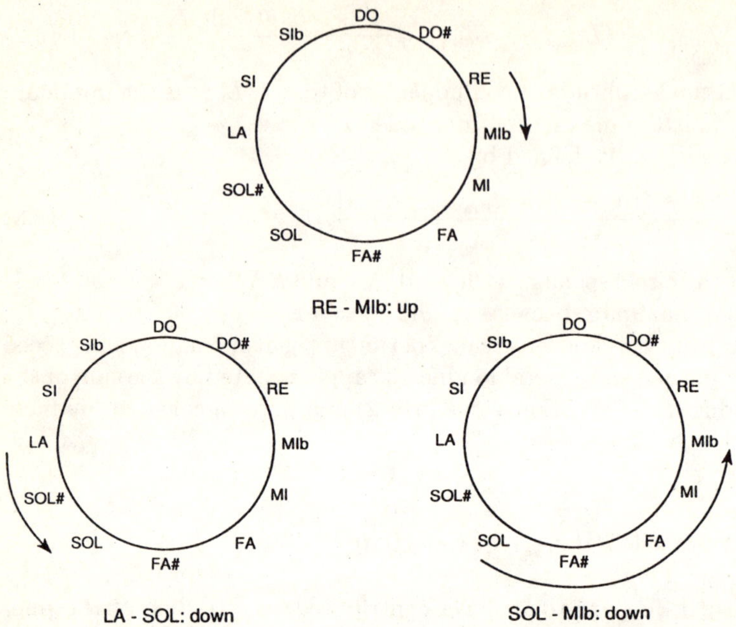

Fig. 3.3. Isotropic height assignment and the principle of proximity

of the nearby positions in the toneness circle. Tritone intervals have an equal distance in both the upward or downward direction so that the pitch system cannot rely on nearby effects. For one listener a tritone will go up, while for the other it will go down. According to Deutsch, listeners seem to be individually consistent about the perceived height direction of the tritones but not intersubjectively consistent: the same interval is heard as up by half of the audience and down by the other half. The paradoxical phenomenon is relevant for both trained and untrained listeners.[2]

3. The alternation of tritone intervals with other intervals corresponds with an on/off switch of proximity effects. This might produce strange effects. The sequence **MI-LA♯-DO-FA♯-LA-RE♯-DO♯-SOL** contains four pairs of tritones. According to Deutsch, the perception of height in this sequence corresponds with the individual height profiles for tritones. One person will hear this as notated in Fig. 3.4 (upper part), the other one as in Fig. 3.4 (lower part).[3]

[2] The dichotomical basis of many of the experimental setups can be criticized. A request to answer either "up" or "down" might force the listener to make choices that would otherwise be judged in a more nuanced way. Our own (informal) experiments with skilled listeners show that they often hear the tritone interval to go both up and down, just as they actually hear more than one tone in the Shepard-tone (Sect. 3.6).

[3] Having replicated this setup, we found that the same person can hear both sequences. Much depends on the will of the listener to hear the first tritone go

Fig. 3.4. Ambiguous Shepard-tone sequences (Adapted from [3.2])

4. Shepard-tones are probably perceived from within a framework of speech perception. A recent study by *Deutsch* et al. [3.3, 6] suggests a correlation between the perception of isotropic height and speech. They found that the highest limit of the octave band containing the highest proportion of fundamentals bears some relationship to the highest position on the toneness circle. The results so far have been based on a small population with different speech characteristics and further research is needed to point out whether these findings can be generalized.

3.5 The Shepard-Illusion

This illusion was first mentioned by *Shepard* [3.15]. It refers to an illusion of absolute height, while only toneness (that is: relative or isotropic height within the octave) is given. The illusion occurs when chromatic sequences (or glissandi) of Shepard-tones of one octave are repeated.

Figure 3.5 shows the illusion graphically. The full line shows the shift of one octave component in function of time. After one octave, the glissando is repeated. Although the glissando does not go beyond the octave, one hears a pitch which is higher than the octave. But the pitch does not raise infinitely (as the name of this illusion wrongly suggests). At a certain moment, the listener gets the impression that the height was already attained before. *Shepard* [Ref. 3.15, p. 2349]:

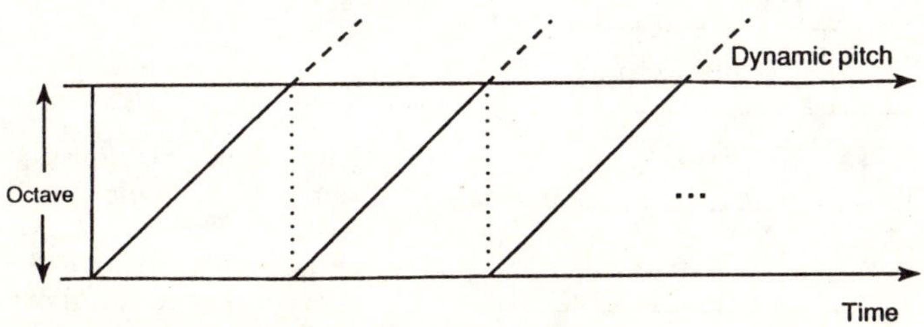

Fig. 3.5. The perception of dynamic pitch

up or down. This can easily be influenced by playing a sequence with piano tones. The sequence with Shepard-tones is then accordingly heard.

> Toward the end of the sequence, some of the listeners became puzzled by the fact that the tones (which clearly had been going up for so long) did not really seem to be getting much "higher". Other listeners, however, did not notice this stationarity of height. Indeed, these latter subjects were astonished to learn that the sequence was cyclic rather than monotonic and that it in fact repeatedly returned to precisely the tone with which it had begun.

The effect of hearing a pitch that was previously heard during the sequence can be called a perceptual *catastrophe*: it is a salient effect of the global perceptual behavior that is caused by small changes in the stimulus.

It is perhaps interesting to note that the illusion depends on the speed of the cycling. This was illustrated by *Risset* [3.14] in an example where the cycling speed was decreased. At the beginning one hears the cycling and no rising pitch but when the glissando slows down the raising pitch emerges. This shows that dynamic effects may play a role in the generation of the pitch percept. Research by *Nakajima* et al. [3.11, 12] supports the view that the

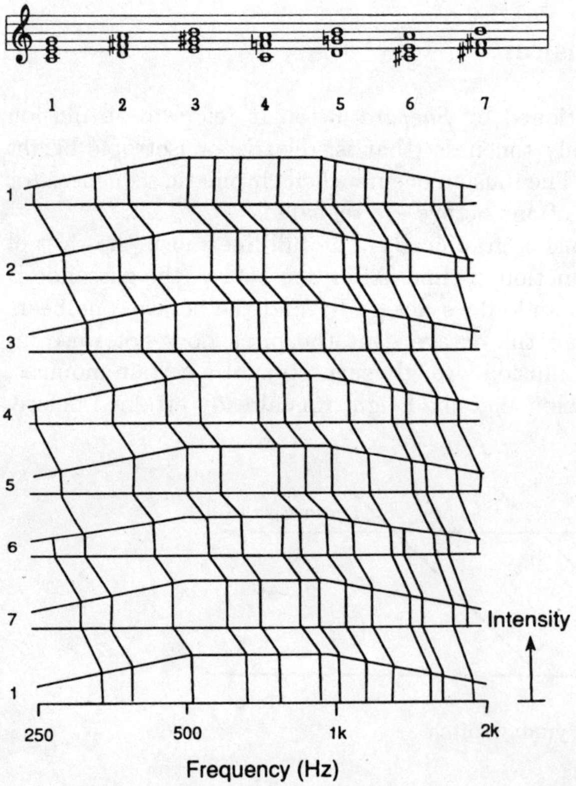

Fig. 3.6. Demonstration of dynamic pitch with Shepard-chords (Based on [3.12]). The spectrum of each chord is labeled in the graph by the number which corresponds to the chord

pitch percept is produced by a temporal dynamic effect: when components are moving in the same direction they contribute to the perception of a dynamic pitch. This refers to the Gestalt principle of common fate – probably an effect of some perceptive principle rather than a sensorial one.

Figure 3.6 shows how the dynamic pitch can be obtained from chord sequences. People are often impressed when they hear the difference between a piano interpretation and a Shepard-tone interpretation. The latter indeed seems to contain pitches which sound higher than the actual notes in the score. The common fate of spectral components is apparent at the chords 3 and 4 and at the cycle point (from chord 7 to chord 1). It might be necessary to adapt the speed of the chords played. If they are played too fast the effect of dynamic pitch might disappear.

3.6 Ambiguous Stimuli

Another interesting effect is produced in the so-called *Risset-illusion.* In this illusion, a tone with frequencies of 49, 102, 211, 435, 896, 1843, and 3788 Hz is followed by the same tone transposed one octave higher (for example by doubling the speed of the tape – the frequencies are then 98, 204, 422, 870, 1792, 3686, and 7576 Hz). For most listeners, the second tone sounds lower than the first one. Some listeners hear a tone going up and down at the same time.

Figure 3.7 shows the effect graphically. The horizontal axis gives the frequency, the vertical axis the amplitude. The long arrows indicate the shift of one octave, while the short arrows show the perceived effect. The effect is explained by the fact that the frequency components fall in the sensitive zone of the auditory system (between 500 and 2000 Hz). The common fate of

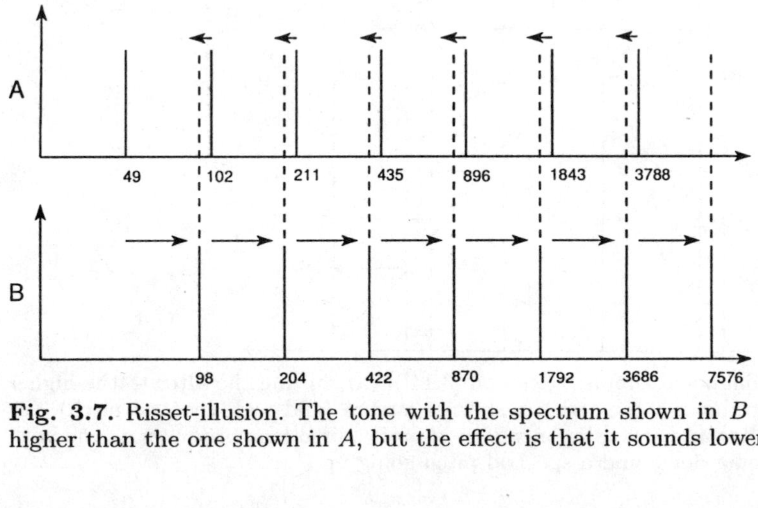

Fig. 3.7. Risset-illusion. The tone with the spectrum shown in B is one octave higher than the one shown in A, but the effect is that it sounds lower

the spectral components, in combination with tone fusion, creates the effect of going down.

Risset [3.14] makes a distinction between the spectral and tonal height ("hauteur spectrale" and "hauteur tonale") of tones. Listeners, when confronted with a complex tone containing the frequencies of 1800, 2000, and 2200 Hz, hear a high pitch (the spectral or timbral pitch) instead of the low pitch (the tonal pitch, fused pitch or residual pitch). He also observed that persons who perceive the low pitch also perceive the high spectral pitch. Musicians seem to be skilled in hearing the low pitch and they seem to rely on this pitch to solve perceptual ambiguities.

Perception can often be influenced on a voluntary basis and the listener can prepare himself to hear the low or high pitch – he often hears both. Of course, the preference for low or high depends on the stimulus as well. With the help of computer synthesis, Risset created ambiguous stimuli in which the tension between spectral tone and residual tone has been exploited.

Starting from Shepard-tones, two varying parameters were introduced. One parameter shifts all the octave components up or down, the effect is a change in low pitch perception. The other parameter shifts the filter shape up or down as in Fig. 3.8a. When the octave components are kept constant, the effect is a change in high pitch perception. Shifting the filter to the higher

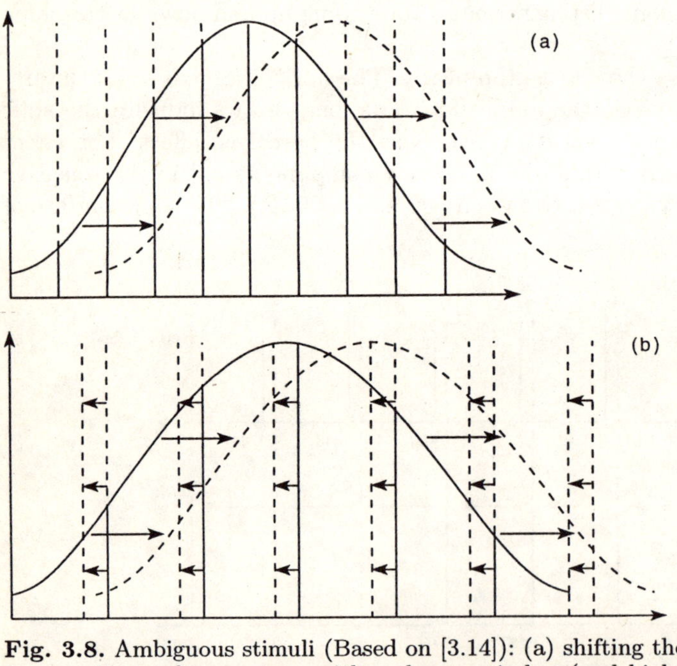

Fig. 3.8. Ambiguous stimuli (Based on [3.14]): (a) shifting the filter to the higher frequencies produces a tone with a sharper timbre (and high spectral pitch), (b) moving the filter and the components in opposite directions produces a tone with a low pitch going down and a spectral pitch going up

frequency regions shifts the energy to the high register and the sound becomes sharper [3.18, 19]. Ambiguous effects were obtained when the shift of the octave components went into the opposite direction of the shift of the filter (Fig. 3.8b). For example, when the octave components go down and the filter goes up, one hears a low pitch that goes down and a high one that goes up. The perception system is fooled in such a way that often choices must be made.

Such experiments lend support to the view that the perception of low pitch and high pitch, with the associated attributes of height and toneness, are determined by concurrent auditory processes. The two channel pitch perception theory by *van Noorden* [3.17] provides a plausible explanation on a physiological basis. The theory relies on the dual coding of the acoustical signal, (a) as the excitation of neurons on particular places along the basilar membrane and (b) as the temporal excitation of neural spikes. Place coding would account for the perception of sharpness and high (spectral) tone, while time coding would account for the perception of low pitch.

3.7 Conclusion

During the seventies and eighties, experiments with Shepard-tones have contributed to the decline of the attribute theory – of which the two-component theory is a cornerstone. The study of illusions has come up with cues, such as the distinction between spectral and residual pitch and the notion of dynamic pitch. As a result of the compelling auditory demonstrations with Shepard-tones and ambiguous stimuli, pitch is now generally regarded as a concept that emerges from auditory information processing. Attributes, such as height, toneness and dynamics, are considered emergent properties of an underlying level. They are determined by Gestalt formation processes that operate in a complex dynamics. Concurrency between these processes is sometimes apparent in illusions and further study of the system dynamics will be needed to obtain a thorough understanding of the effects of common fate. To summarize, pitch perception now tends to be seen from two viewpoints:

- a phenomenological level at which "paradoxes" and "illusions" play a central role as critical determinants of perceptual constraints, and
- a sub-phenomenological level at which illusion and paradoxes are explained and understood by models of auditory information processing.

Illusions are often self-contained and do not really depend on a musical context. Nevertheless, their analysis defines boundaries for the study of tone semantics in that some aspects can be shown to be relevant, while others are less relevant. The notion of low pitch is relevant because it can be related to the concept of "tonal gravity" of chords. Also the idea that perception is based on concurrent processes is important. It offers suggestions for an analysis of the perception system in terms of dynamic systems.

In the subsequent chapters, Shepard-tones will play a central role – not because of their illusion generating capacity – but because they allow a considerable *reduction* of data. Shepard-chords (chords made of Shepard-tones) have no inversions and can therefore be considered to be the expression of the purest possible form of toneness. Also the phenomenon of dynamic pitch is important, but its relevance, and possibly also the mechanisms by which it is generated, can be discarded in the present study.

4. Defining the Framework

This chapter describes the general framework for a computer model in which a schema theory for tone center perception is developed. It provides a basis for modelling the emergence of the pitch phenomena discussed in the previous chapter. The framework is based on a causative connection between different representational categories: signals, images, schemata. The knowledge structure of the schemata which come out of the model can be compared with the structures of mental representations. This provides a well-defined paradigm for the study of music cognition.

4.1 The Computer Model

The computer model is shown in Fig. 4.1. The task of this model is to recognize and interpret tone centers in a musical signal. It consists of two modules: a perception module and a cognition module.

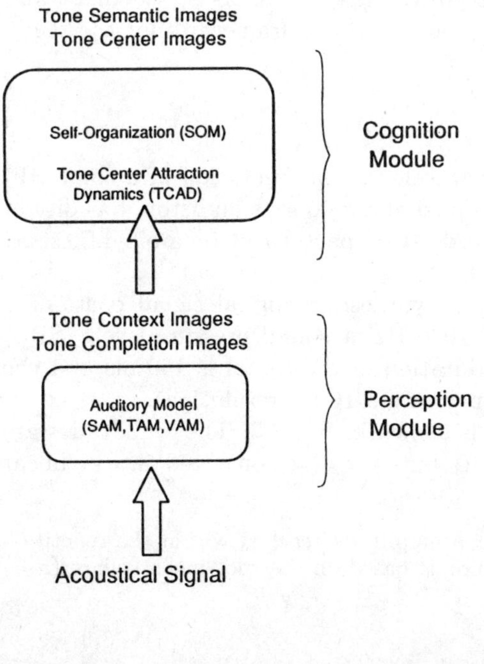

Fig. 4.1. The computer model consists of a perception module and a cognition module. The input is an acoustical musical signal. Auditory images are produced at different stages of the model

- The perception module contains an auditory model which transforms the musical acoustical signal into a flow of auditory images. Two important types of images are tone *completion* images and tone *context* images. Context images convey context-sensitive information about a recent past. The auditory models used in this book are called SAM, TAM, and VAM. They are reviewed in Chap. 5.
- The image flow generated by the perception module is further processed by a cognition module. This module comprises two types of dynamics:
  - *Self-organization*: this is a data-driven long-term learning process by which neurons adapt to the environment. Self-organization leads to a stable tone perception schema. A model for self-organization as a process of learning is described in Chap. 6. The computer simulations are discussed in Chaps. 7–8.
  - *Tone center attraction*: this is a schema-driven short-term control process by which signals are recognized and interpreted. The dynamics of tone center attraction is a form of associative self-organization. Its principles are described in Chap. 9. The computer simulations are evaluated in Chap. 10.

The tone *center* images reflect the schema's emergent representational states. These are used as a schema for tone center recognition and interpretation. The *semantic* images, which are based on this process, are discussed in Chap. 10.

4.2 Representational Categories

In modelling perception and cognition it makes sense to distinguish between four types of "representational categories": signals, images, schemata and mental representations. These categories provide a framework for the study of tone semantics.

4.2.1 Signals

A *signal* refers to the acoustical or waveform representation of a sound. In the computer model, signals are sampled at 20000 sa/s and stored as digital signals. Signals are taken from records (Compact Disk) or are synthesized with a sound compiler.

Figure 4.2 gives an example of a synthesized digital signal containing frequencies of 600 Hz, 800 Hz, and 1000 Hz at sound pressure levels (SPL) of 60 dB, 55 dB, and 50 dB. The duration of the signal is 100 ms and the amplitude is in linear representation (using a 16-bit resolution).

The signal generating program is shown in Fig. 4.3. The amplitudes are specified as peak amplitudes[1] in dB but are first converted into a linear representation according to

[1] The peak amplitude is the maximum amplitude reached within the repetition period. Another measure of amplitude is based on the root mean square (rms)

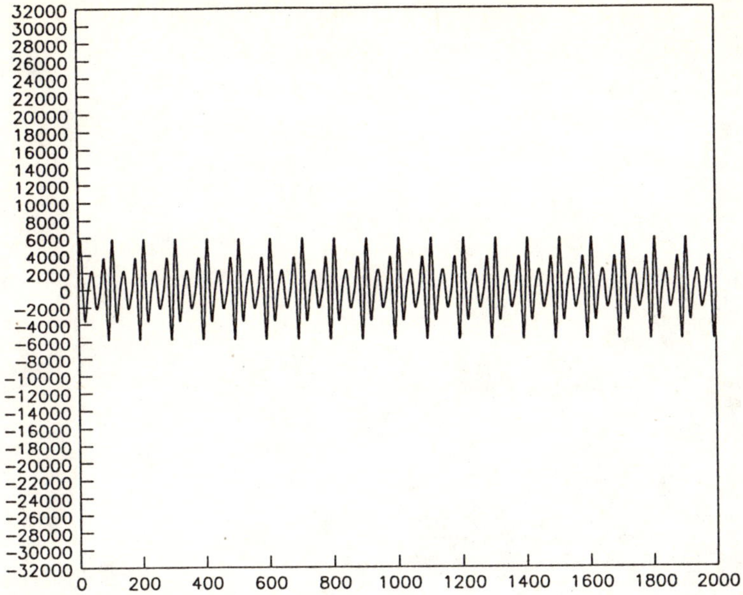

Fig. 4.2. Digital representation of a complex tone in linear amplitude

$$\mathrm{ALIN} = \frac{10^{\mathrm{ALOG}/20}}{10^{\mathrm{ALOG}_{\mathrm{max}}/20}}. \tag{4.1}$$

The maximum amplitude ($\mathrm{ALOG}_{\mathrm{max}}$) in the model is 80 dB, so that the numerator becomes 10000. ALOG is the amplitude expressed in dB (logarithmic value). Since the amplitudes have to be represented by integers (16 bits) in stead of reals, the ALIN values are multiplied by the maximum 16 bit integer value: $\mathrm{ALIN}_{\mathrm{max}} = 32727$.

The peak amplitude of the compound signal can be found by translating the sum of the linear amplitudes back into logarithmic values, using

$$\mathrm{ALOG} = 20 \log_{10} \left(\frac{\mathrm{ALIN}\sum}{\mathrm{ALIN}_{\mathrm{max}}} \right) + \mathrm{ALOG}_{\mathrm{max}}, \tag{4.2}$$

where ALIN $\sum$ contains the sum of all linear amplitudes. In this example, the peak amplitude is 65.8 dB.

In music research, it is sometimes necessary to perform classical digital signal processing operations on musical signals. Figure 4.4 shows a 4096-point FFT based power spectrum analysis, using a Hamming window.[2] The frequency is shown on a linear scale from 500 Hz to 2000 Hz. The peaks in this spectrum are found by a simple peak detection algorithm which first sets

value of the signal. This value is proportional to its energy content. The ratio of the peak amplitude to the rms amplitude is 1.414 (± 3 dB).

[2] Because of the window, small deviations may occur if the frequency is not an exact multiple of a bin (the sampling rate divided by the window length).

```
#include <stdio.h>
#include <math.h>
#include <fcntl.h>

#define PI 3.141592654
#define T1 600.0
#define T2 800.0
#define T3 1000.0
#define DB1 60.0
#define DB2 55.0
#define DB3 50.0
#define DBMAX 80.0
#define TIME 2000
#define SA 20000.0
#define MAXINT 32767.0

main()
{
double sn, scale1, logtotscale, totscale, scale2, scale3, scalemax;
int i, fhout, ret;
short snout;

fhout=creat("H3.F2.D",00660);
scalemax=pow(10.0,DBMAX/20.0);

scale1=(pow(10.0,DB1/20.0)/scalemax) * MAXINT;
scale2=(pow(10.0,DB2/20.0)/scalemax) * MAXINT;
scale3=(pow(10.0,DB3/20.0)/scalemax) * MAXINT;
totscale=scale1+scale2+scale3;
logtotscale=(20 * log10(totscale/(double)MAXINT)) + (20 *log10((double) scalemax));
fprintf (stderr, "%.2lf + %.2lf + %.2lf = %.2lf (=%.2lf)\n",
         scale1, scale2, scale3, totscale, logtotscale);

for (i=0; i<TIME; i++){
 sn=scale1 * sin(2.0 * PI * T1 * i /SA);
 sn+=scale2 * sin(2.0 * PI * T2 * i /SA);
 sn+=scale3 * sin(2.0 * PI * T3 * i /SA);
 snout=(short) (sn);
 ret=write (fhout,&snout, 2);
 };
close(fhout);
}
```

Fig. 4.3. Example of a computer program written in C, for generating a complex tone

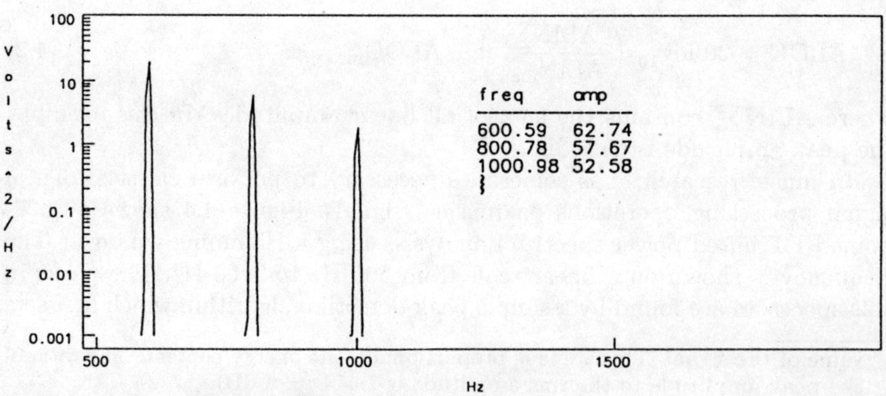

Fig. 4.4. Power spectrum analysis of a complex tone

all values below a given reference value to zero. This divides the array into sequences for which the maximum value can be computed.

A list of frequency-amplitude pairs is obtained by taking the corresponding frequencies of the peaks in the array and transforming the linear amplitude values into dB-values. The amplitude values of a power spectrum represent power (not pressure), for which (4.3) must be used.

$$\mathrm{ALOG} = 10 \log_{10} \left(\frac{\mathrm{ALIN}}{\mathrm{ALIN_{max}}} \right) . \tag{4.3}$$

Lists of frequency-amplitude pairs can be used as input to the auditory model TAM (Sect. 5.4). Figure 4.5 contains a list of frequency-amplitude pairs describing the spectrum of a Shepard-tone **DO**. A small computer program calculates this list of frequency-amplitude pairs using (3.1). The range between the highest and lowest peak amplitude is 40 dB. To avoid too much difference in amplitude in the sensitive region, the amplitudes of the frequencies between 500 Hz and 1000 Hz have been cut off (clipped) at 70 dB.

The synthesis of such a tone can be efficiently realized with a sound compiler. Figure 4.6 shows a MUSIC5 code with corresponding graphical aid. Once compiled, it generates a signal which is identical to the one shown in Fig. 4.3. Lists of frequency-amplitude pairs can easily be inserted into the program code of the sound compiler. In this book, the synthesized sounds have been produced with MUSIC5 [4.8] and CSOUND [4.14].

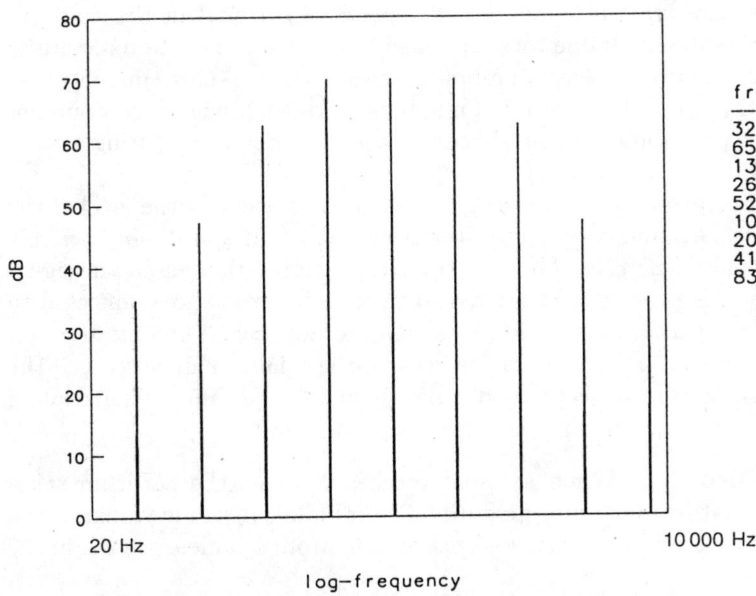

freq	dB
32.70	34.77
65.40	47.27
130.80	62.73
261.60	70.00
523.20	70.00
1046.40	70.00
2092.80	62.73
4185.60	47.27
8371.20	34.77

Fig. 4.5. Spectral representation of the Shepard-tone of **DO**

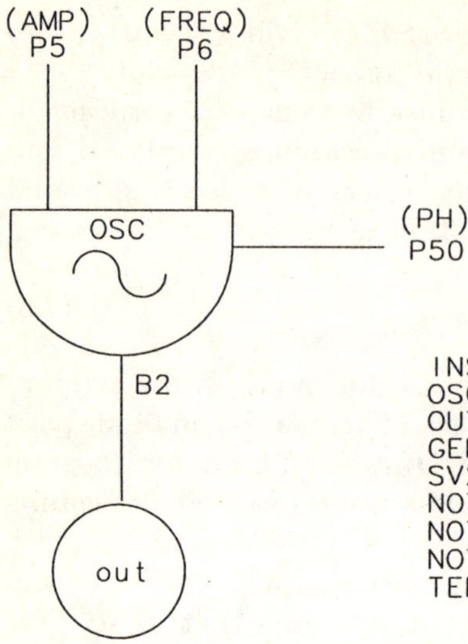


```
INS 0 1;
OSC P5 P6 B2 F1 P50;
OUT B2 B1;
GEN 0 2 1 512 1 1;
SV2 0 20 2 205 6;
NOT 0.00 1 1 60 600 ;
NOT 0.00 1 1 55 800 ;
NOT 0.00 1 1 50 1000 ;
TER 1;
```

Fig. 4.6. The code specifies a simple sinusoidal oscillator with amplitude (AMP) and frequency (FREQ) as input, the phase (PH) is zero. The variables for AMP and FREQ (P5 and P6) are instantiated in those lines that start with "NOT". The fifth field (P5) is the amplitude, the sixth field (P6) is the frequency

4.2.2 Images

An *image* or *perceptive map* represents aspects of a signal in the auditory system.[3] In the present framework, an image is assumed to contain a state or snapshot of the neural activity in a brain region during a short time interval. It is modelled as an ordered array of numbers (a vector), where each number represents the probability of neural firing of a group of neurons during a short time interval.

The most complete auditory image is assumed to occur at the level of the *auditory nerve*. An auditory nerve *fiber* is an *axon* of a *spiral ganglion cell*, which is driven by *hair cells*. The cells transform information about the signal activating the receptor cells connected to them. The axons are connected to the *cochlear nucleus*, where auditory processing becomes differentiated and more specialized [4.15]. (Some of its cells are specialized in detecting the onset.) The following aspects should be taken into account when dealing with auditory images:

- **Spatial Encoding**. When a sound reaches the ear, the eardrum takes over the variations in sound pressure. The middle ear beans transmit the vibration to the cochlea and a sophisticated hydromechanical system in the

[3] Readers interested in the details of auditory information processing are referred to handbooks such as [4.5, 6, 9]. Appendix B contains a brief summary of the essentials of physiological acoustics.

cochlea then converts the vibration into electrochemical pulses. Depending on the temporal pattern of the signal, a traveling wave pattern is generated in the cochlear partition which produces a characteristic spatial configuration [4.4]: low frequencies are encoded near the apex of the cochlea, while high frequencies are encoded near the stapes.

- **Temporal Encoding**. Due to the membrane properties of the spiral ganglion cell, the graded activity of the inner hair cells generate all-or-non activity (action potentials or *spikes*) in the auditory nerve fibers that innervate these cells. Neurons are in general incapable of generating more than 500–1000 spikes per second. As a result of this, the neural response to an acoustical stimulus will reflect the envelope of the fine temporal structure of the stimulating wave form. This capacity of the neurons to *synchronize* has been shown to be very important in the perception of low pitch, and timbre [4.7]. We assume that it plays a central role in tone center perception as well. Some images used in the model are therefore based on an analysis of the temporal encoding patterns.
- **Multiple Images**. Since the auditory system encodes information in multiple ways, it makes sense to distinguish between different types of images. For example, *Brown* [4.2] makes a distinction between auditory nerve images, onset images, offset images, residue images, frequency transition images, etc. In tone center recognition, relevant images are: auditory nerve images, autocorrelation images, tone completion images, tone context images, tone semantics images, etc. In rhythm research, relevant images are provided by the rhythmogram [4.13]. In timbre research, envelope spectrum images or synchrony spectrum images can be used [4.3]. The present model assumes that these images convey different aspects of the signal at the auditory level. Their exploration and combination is perhaps one of the most fascinating aspects of modelling.
- **Temporal Resolution**. Neurophysiological data suggest that the brain uses distinct temporal resolutions at different levels of the auditory system. For example, the temporal resolution of the auditory nerve fiber has been found to be about 10 times higher than the temporal resolution in the inferior colliculus and about 100 times higher than the neurons of the auditory cortex [Ref. 4.10, p. 355]. The idea of having multiple temporal resolutions has been adapted to the model. For example, the images of the periphery are updated faster than those of the center levels. In the auditory model VAM (Sect. 5.5), the auditory nerve images are updated every 0.4 ms, the completion and context images only every 10 ms. The semantic images, which provide the information of the tone center analysis at a cognitive level, are updated every 100 ms.
- **Images-out-of-Time and Images-in-Time**. *Images-in-time* are images that are formed by the auditory system at defined intervals. The temporal order of their appearance reflects the flow of information. *Images-out-of-time* are abstracted from images-in-time: they do not reflect the temporal

order of processing and do not contain any information about the duration of the auditory object, nor about the immediately preceding past. Such images are used in the model as shorthand images – as in Chap. 7 and Chap. 9.
- **Context Images**. It is assumed that much of the context-sensitivity of the brain is actually due to the capacity to integrate neural discharges. As such, the differences in temporal resolution in the periphery and the center of the brain can be interpreted as reflecting differences in context-sensitivity. Images obtained by large integration (typically a few seconds) are called *context images*. They contain information about a preceding past – the musical context.

4.2.3 Schemata

Bregman [Ref. 4.1, p. 401] has recently defined a schema as a control structure in the human brain that is sensitive to some frequently occurring pattern, either in the environment, in ourselves, or in how the two interact. In the current framework, a schema is a categorical information structure which reflects the learned functional organization of neurons in its response structure. As a control structure, it performs activity to adapt itself and to guide perception. The following aspects should be taken into account when dealing with schemata:

- **Functional Organization**. Neurons that belong to a certain area can have (or develop) a particular functionality or response property depending on a given stimulus. There is evidence that neuronal functions of different nuclei in the auditory brain are ordered according to a specific axis of frequency. These so-called *tonotopic maps* correspond to the spatial coding of frequency along the basilar membrane. In general, the organization of the cerebral cortex is such that neuronal cells with common specificities are grouped together and are separated from cells with other specificities. Functional organization is the basic feature of any self-organizing model.
- **Multiple Levels**. As *Zeki* [4.16] shows for vision, connections in the cortex are commonly of the "like-with-like" type, one group of specific cells in one area connecting with their counterparts in another area. There is indeed evidence for a number of different types of maps, apart from the tonotopic or cochleotopic representation [4.10–12]. In the model, it is assumed that tone center recognition is based on a cortical map which is specialized in pitch perception.
- **Level of Integration**. Maps that belong to a certain sub-modality may also differ with respect to their level of integration. Some maps are specialized in low level responses, while others operate on a higher level. In auditory processing one may assume that the cochleotopic maps, which represent the cochlea much like the brain area V1 represents the retina,

are low level maps. Maps, such as those for form or color vision rely on low-level maps but they are "high level" because they respond to more complex features of the signal. In the model, the level of integration is defined by a preprocessor which extracts specific features from the acoustical signal. As such, the self-organizing map for tone center recognition is assumed to be located at a high (cognitive) level.

- **Long-Term Data-Driven Learning and Short-Term Schema-driven Control**. Schemata are multifunctional and it is possible to distinguish between long-term data-driven activity and short-term schema-driven activity. In the present model, both processes are separated. Adaptation to the environment is seen as a long-term process which takes several years. It is data-driven because no pre-defined knowledge is needed in adapting to the environment. Short-term schema-driven control is a short-term activity (3–5 s) which is responsible for recognition and interpretation. It relies on pre-defined knowledge which is contained in the schema.

The above distinctions give but a rough approach to the framework of the current study. More details will be given in the subsequent chapters. One problem of immediate relevance here concerns the relationship between image and schema. One may conceive of this relationship as follows. The responses of the model are always considered as images. In that sense, the response of a schema to an image is also an image. But the schema has an underlying response and control structure which is more persistent than images. The structure contained in a schema is long-term, while the information contained in an image is short-term. The latter is just a snapshot of an information flow.

4.2.4 Mental Representations

Mental representations are knowledge structures that are supposed to be used in performing specific tasks (Sect. 2.4). Mental representations are derived from psychological testing and they refer to a "mental" world, rather than a "physiological" or "brain" world. By using techniques of multi-dimensional scaling, it is possible to depict the data of the tests as mental spaces (Fig. 2.3).

4.3 Conclusion

The aim of this chapter has been to provide a general framework, the details of which will be worked out in subsequent chapters.

In summary, one of the aims of modelling perception and cognition is to show that the representational categories are somehow causally related to each other. Signals are transformed into images, and images organize into schemata and are controlled by these schemata.

By looking for correlations between responses produced by a model, and the data gathered by cognitive psychologists, one may try to relate the response structure of the schema to the space of mental representations. The present study aims to show that the mental structures for tone center perception can be completely understood in terms of causal relations between signals, images and schemata.

A final remark concerns the type of images used in this study. The images all have a *frame-based* nature in that their representation relies upon analysed frames (short segments of information). *Streams* or *continuous-time* images [4.1, 2] are not used in this approach.

5. Auditory Models of Pitch Perception

This chapter gives an introduction to auditory models and pitch perception. Three different auditory models are introduced: a simple model, a place-model, and a place-time model. Only those aspects are reviewed that are relevant for the schema theory.

5.1 The Missing Fundamental

When we listen to the tone of a piano we hear a pitch that corresponds to the fundamental of the harmonic complex. When the fundamental and some of the first harmonics are filtered out, we hear a tone with a slightly different timbre but with the same pitch as the fundamental. This phenomenon is known as the perception of the missing fundamental. The perceived pitch is called the *low pitch*, *virtual pitch* or *residue pitch*. The latter term suggests that the pitch percept is generated by its spectral residue.[1]

The implications of low pitch perception are straightforward in daily life. When making a phone call, the pitch of a male voice is often the pitch of the missing fundamental. The limited bandwidth (e.g., from 300 to 3400 Hz) at which speech signals are transmitted filters out low and high frequencies, but the auditory system reconstructs the correct pitch. Similarly, cheap transistor radios filter out low frequencies, but we hear the low tones without effort. Our auditory system seems to behave as a pattern completion device; the missing fundamental is reconstructed from the residue.

Terhardt [5.16–18] and *Parncutt* [5.3–5] argue that a model of virtual pitch perception is fundamental to a psychoacoustical foundation of harmony. Their approach is grounded on the spatial encoding properties of the auditory system. A chord is assumed to support low pitches, resulting from the spectral relations in the ear's excitation patterns. By using spectral harmonic sieves and subharmonic templates, the computation of the virtual pitch patterns of chords is very straightforward. The similarity relationships between these virtual pitch patterns seem to correspond well with musical intuition and psychological data [5.4].

[1] Strictly speaking, a residue pitch is a virtual pitch that does not correspond to a spectral component of the tone. Most virtual pitches correspond to spectral components.

Most important for music is that the process of low-pitch perception by completion is not just limited to performance under bad conditions. Listening to a chord played by an acoustic piano will produce images that determine harmonic progressions. Chords generate completion patterns that correspond to basic properties of the chord as a whole. The pattern represents the "tone gravity" of a chord and is related to Stumpf's concept of "Verschmelzung" (tone fusion) and Rameau's concept of the "Basse Fondamentale" (fundamental bass). One may say that contrary to the perception of implied fundamentals, the root of the chord is not necessarily heard (in the chord **DO-MI-SOL** the root is the **DO** which is two octaves lower), yet it plays an important role in music.

5.2 Auditory Models

The schema theory dealt with in this book has been developed over several years and during its development, three different auditory models have been used. The acronyms are:

- SAM, a *Simple Auditory Model*, based on the reduction of toneness,
- TAM, the *Terhardt Auditory Model*, based on the auditory model by *Terhardt* et al. [5.17], and
- VAM, the auditory model by *Van Immerseel* and *Martens* [5.4].

SAM is a simple model, developed in 1986 to demonstrate the emergence of a tone perception schema by self-organization [5.7–11]. The model has since been adapted to the one used by *Parncutt* [5.13]. Both TAM and VAM have been developed by researchers working in speech recognition departments. The differences between TAM and VAM are above all exemplary for a certain "style" in which to approach the subject. TAM is an auditory model based on the spatial encoding principles of the auditory system. It has been implemented using *functional* principles rather than digital filtering theory. In contrast, VAM relies on spatio-temporal encoding principles and is implemented with digital filters.

The conceptual differences between these models (both in implementation and concept) are great but their use in the tone semantics model have led to consistent and often surprisingly similar results. The multiplicity of approaches is a reality of scientific research, and for that reason, all three models are used here. In principle, all results could have been obtained with the most recently used model, nl. VAM.

5.3 SAM: A Simple Model

Due to our limited computing resources in 1986, data reduction was one of our main concerns, but experiments with the Shepard-tones suggested a way to reduce residue images to their toneness counterpart – yielding a pattern of only 12 dimensions.

As pointed out in Sect. 3.3, the subjective effect of the perception of Shepard-tones indeed suggests a reduction of the whole audible frequency range to one octave. A chord and its inversions can thus be represented by a prototype chord over one octave. (Height is an important factor in harmony, but it is not essential to the perception of tone center.)

Shepard-tones have been used in psychological tests and the testing of the computer model could therefore rely on the same type of data. Of course, listeners were not trained with pure Shepard-music and at this point the situation differs from the "real" situation. Pure toneness patterns, however, suffice to demonstrate some of the basic principles in tone semantics. Working with the simplified models is justified as long as it entails a straightforward extension to "real-world" simulations.

5.3.1 SAM – The Acoustical Representation

The reduction of the frequency range is based on the frequency range of perceived Shepard-tones, which goes from f to 2f. In the equal-tempered chromatic tone scale this range is divided into 12 equal frequency steps. Then it becomes possible to represent the Shepard-tone signal by a 12-dimensional pattern and name the frequency components (that rest further unspecified) as notes: **DO**, **DO♯**, **RE**, **MI♭**, **MI**, etc...

The *S-representation* (or Simple representation) of the Shepard-tone **DO** is given by:

1 0 0 0 0 0 0 0 0 0 0 0 **DO**

The last element of the pattern is the label (which is optional). Shepard-chords are represented by patterns such as:

1 0 0 1 0 0 0 1 0 0 0 0 **Cm**
1 0 0 1 0 0 1 0 0 1 0 0 **Co7**

The chord **Cm** comprises the tones **DO**, **MI♭**, and **SOL**, whereas the chord **Co7** comprises the tones **DO**, **MI♭**, **FA♯**, **LA**.

Shepard-music is represented by a sequence of S(imple)-patterns (possibly without labels) where each S-pattern is interpreted as a sample of a spectral pattern over a certain period of time. For other purposes, the patterns can be interpreted as patterns "out of time", which yields an important reduction of data.[2]

[2] For the distinction between patterns "in time" and patterns "out of time", see Chaps. 7–8.

5.3.2 SAM – The Synthetic Part

The manipulation of the acoustical representation must be accounted for in the auditory model as well. Given an S-pattern, abstraction can indeed be made of the frequency analysis part of the auditory model because the components can be seen as resolved spectral components. In terms of the framework sketched in Sect. 4.2, the S-pattern serves as the representation of the signal *and* as a spectral image. In the latter case the components stand for spectral components that are resolved by the hydrodynamic-based filtering in the cochlea.

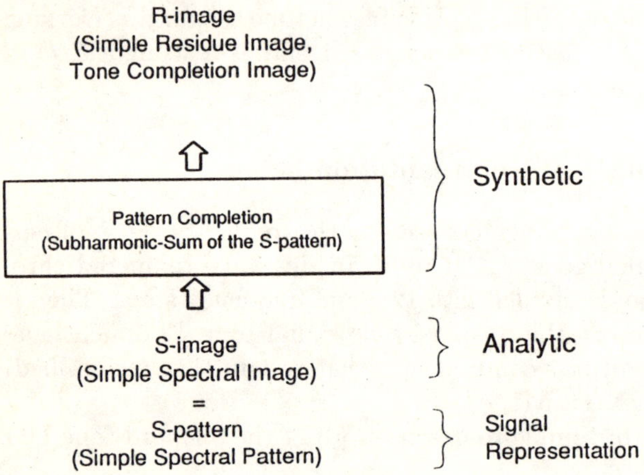

Fig. 5.1. Overview of SAM

Figure 5.1 gives an overview of the model. The synthetic part comprises the completion process. The resulting image (called *tone completion image*, *simple residue image* or *R-image*) is calculated as the subharmonic-sum of the S-image, that is: the juxtaposition of the weighted subharmonics of each component in the S-image. Equation (5.1) gives a concise mathematical formulation.

$$R_i = \sum_{j=0}^{11}(W_j * S_{(i+j)\%12}). \tag{5.1}$$

R_i is the i-th component of the R-image (from 0 to 11) and W_j is the weight which is associated to the interval between components S_i and $S_{(i+j)\%12}$. The octave-reduction is accounted for by the modulo operator "%".

The outcome of the above expression is clarified in Table 5.1, with an example of the major-triad chord **DO-MI-SOL**. The S-image is shown at the top. The R-image is shown below, together with the note names. The weights W associated to the subharmonic intervals are given to the left.

Table 5.1. Subharmonic-Sum Table

S-image:	1	0	0	0	1	0	0	1	0	0	0	0
octave(1.00)	DO	-	-	-	MI	-	-	SOL	-	-	-	-
fifth(0.50)	SOL	-	-	-	-	DO	-	-	-	MI	-	-
maj.third(0.33)	MI	-	-	SOL	-	-	-	-	DO	-	-	-
min.sev.(0.25)	-	-	DO	-	-	-	MI	-	-	SOL	-	-
maj.sec.(0.20)	-	-	MI	-	-	SOL	-	-	-	-	DO	-
min.third(0.10)	-	MI	-	-	SOL	-	-	-	-	DO	-	-
R-image:	1.83	0.10	0.45	0.33	1.10	0.70	0.25	1.00	0.33	0.85	0.20	0.00
Notes:	**DO**	**DO♯**	**RE**	**MI♭**	**MI**	**FA**	**FA♯**	**SOL**	**LA♭**	**LA**	**SI♭**	**SI**

The first eight subharmonics are, in descending order: **DO** (related to the original **DO** by an octave below), **FA** (related to the **DO** by a fifth below), **DO** (octave), **LA♭** (major third), **FA** (fifth), **RE** (minor seventh), **DO** (octave) and **SI♭** (major second below **DO**). Since octave components are most strongly present in this subharmonic series, a fixed weight of 1.0 is associated. The fifth occurs twice in this series and gets the weight of 0.5. The subharmonic intervals of major third, minor seventh, and major second occur only once. They are weighted of 0.33, 0.25 and 0.20. The introduction of the minor third (weight=0.10) is a special case, because the interval does not occur in the first 10 harmonics.[3]

The result of transforming the above S-images of **DO**, **Cm** and **Co7** into R-images is:

[3] The weights have been adapted to be in agreement with [5.13]. According to R. Parncutt, the minor third can support the root because its third and fifth harmonics (the perfect fifth and major third) are octave-equivalent to the root's seventh and third harmonics (minor seventh and perfect fifth). More recently [5.15], this reasoning is not followed anymore since it applies also for other intervals such as the major sixth. Actually, different weights are conceivable but the above set is also related to the one used in TAM. In TAM the weights W of the subharmonics are inverse proportional to the harmonic number ($W_i = 1/i$, for $i = 1$ to 15). In *SAM*, the weights that occur at different octaves are added. The octave-intervals then become $1/1 + 1/2 + 1/4 + 1/8 = 15/8$. Similarly, the total weight of components at fifth-intervals is $1/3 + 1/6 = 1/2$, while the components at major third-intervals is $1/5 + 1/10 = 3/10$. The obtained values are: 15/8 (added octaves), 1/2 (added fifths), 3/10 (added major thirds), 1/7 (only one minor seventh), and 1/9 (only one major second). Normalized (divide by 15/8) this gives: $1.0, 0.27, 0.16, 0.08$, and 0.06 respectively. According to [5.13], an intuitively more reasonable set of values is obtained by raising the numbers to the power of 0.55. One then obtains the values depicted in Table 5.1. The weights approximate the series $1/1, 1/2, 1/3, 1/4, 1/5$.

```
1.00 0.00 0.25 0.00 0.00 0.50 0.00 0.00 0.33 0.10 0.20 0.00 DO
1.60 0.20 0.25 1.33 0.10 0.95 0.00 1.00 0.83 0.35 0.20 0.33 Cm
1.10 0.20 0.58 1.10 0.20 0.75 1.00 0.00 1.08 0.10 0.20 0.83 Co7
```

Since the list contains no indication of time, these images should be considered "images-out-of-time".

5.4 TAM: A Place Model

TAM is a place model – based on the spatial encoding principles of the auditory system. The virtual pitch information is extracted from the spectral-pitch images.

E. Terhardt was the first to show the importance of virtual pitch theory for music research. With the introduction of the subharmonic-sum spectrum algorithm in 1974, TAM became a popular pitch model in music research. Some computer simulations reported in this book rely on the virtual pitch program developed by Terhardt and his collaborators. Readers who are interested in the mathematical details can look at [5.17, 18]. The processing steps of TAM are depicted in Fig. 5.2.

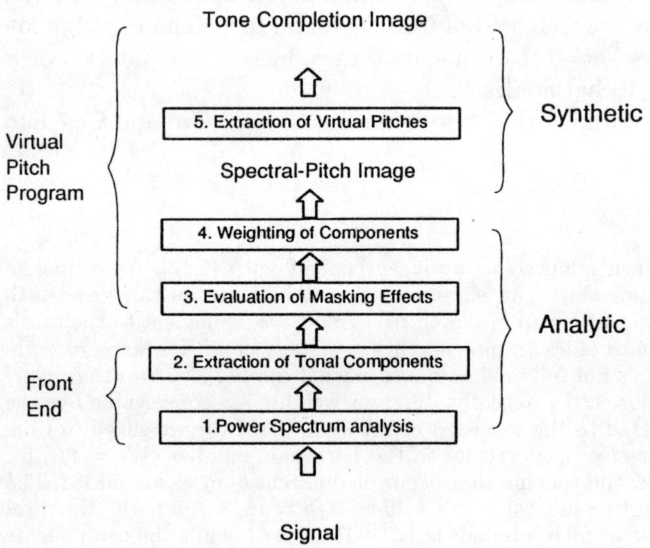

Fig. 5.2. Overview of TAM

5.4.1 TAM – The Analytic Part

The analytic part extracts the spectral-pitch components that are relevant for the calculation of the virtual pitch images. It consists of a power spectrum analysis (step 1) of the signal, out of which the candidates for tonal components are extracted (step 2). The masking effects (step 3) take into account the fact that components may be inaudible or their audibility is reduced due to the presence of other components, as well as the fact that components will be shifted as a result of mutual partial masking. The weighting of the spectral components (step 4) accounts for the available evidence on spectral dominance and loudness effects. The resulting image, the spectral-pitch image, forms the input to the module which extracts the virtual pitch (step 5).

The extraction of tonal components provides a list of frequency-amplitude pairs which is the input to the virtual pitch program.[4] Frequency-amplitude lists provide a shorthand for data reduction, in particular the creation of images-out-of-time (similar to SAM).

5.4.2 TAM – The Synthetic Part

Pitch completion is based on the computation of the subharmonic-sum spectrum (SHSS) of the spectral-pitch images. The prototype SHSS is given by

$$W(f) = \sum_{m=1}^{15} \frac{S\,(m * f)}{m}, \tag{5.2}$$

where $W(f)$ is the summed weight of frequency f, and $S\,(m * f)/m$ is the weight contribution of the m-th harmonic of f. The function $S\,(.)$ outputs the unit value (=1) when the frequency f, multiplied by m, occurs in the resolved spectrum (otherwise $S\,(.) = 0$). Up to 15 harmonics are considered. For example, if the resolved spectral components are 600 Hz, 800 Hz, and 1000 Hz, then the values for virtual pitches at 100 Hz, 200 Hz and 400 Hz are

```
...
W(100) = 1/6 + 1/8 + 1/10 = 0.39
...
W(200) = 1/3 + 1/4 + 1/5 = 0.78
...
W(400) = 1/2 = 0.5
...
```

A complete subharmonic-sum spectrum (SHSS) is obtained when the calculation is done for all frequencies which cover the virtual pitch range (approximately from 50 Hz to 500–800 Hz). A graphical representation is shown in

[4] The front end, containing power spectrum analysis and extraction of tonal components, can be replaced by other techniques for getting frequency-amplitude lists. Sect. 4.2.1 describes how such lists can be obtained from a signal.

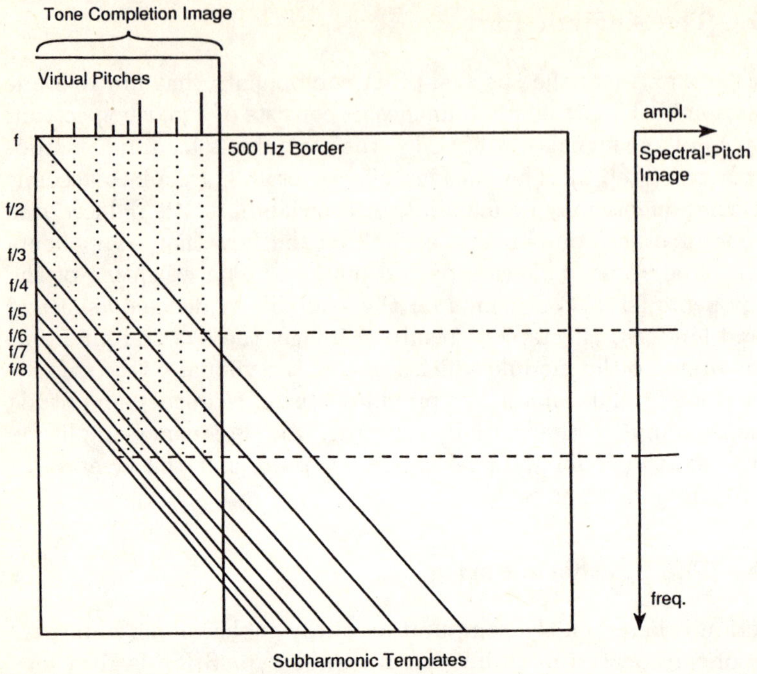

Fig. 5.3. The spectral-pitch image is shown at the right, with frequency pointing down and amplitude pointing to the right. There are two spectral components which are mapped onto virtual pitches at the points where they cross the subharmonic templates. The weights, depending on the subharmonic number, have been added up and 500 Hz is taken to be the border for the occurrence of virtual pitches

Fig. 5.3. The diagonal lines correspond to the subharmonic sieves. The sieves can be regarded as a series of narrow slots spaced at the frequencies of the subharmonics. In practice, these slots extend beyond one frequency. In TAM, a coincidence interval of 8% is used and the weight of the virtual pitch is increased with increasing coincidence.

A further adaptation of the prototype subharmonic-sum algorithm (5.2) concerns the amplitude of the resolved spectral components. In TAM, a high amplitude contributes more to the weight of the virtual pitches than a low amplitude. The contribution of a 60 dB tone at 600 Hz will therefore be slightly more important than the contribution of the 800 Hz and 1000 Hz tone components with a peak amplitude of 55 dB and 50 dB, respectively.

5.4.3 TAM – Examples

Figure 5.4 shows the output of the virtual pitch program to the frequency-amplitude list containing the frequencies 600 Hz, 800 Hz, and 1000 Hz, at 60 dB, 55 dB, and 50 dB (Sect. 4.2.1). The table contains the frequency indications of the *nominal* pitch (NP) and the *true* pitch (TP). TP takes

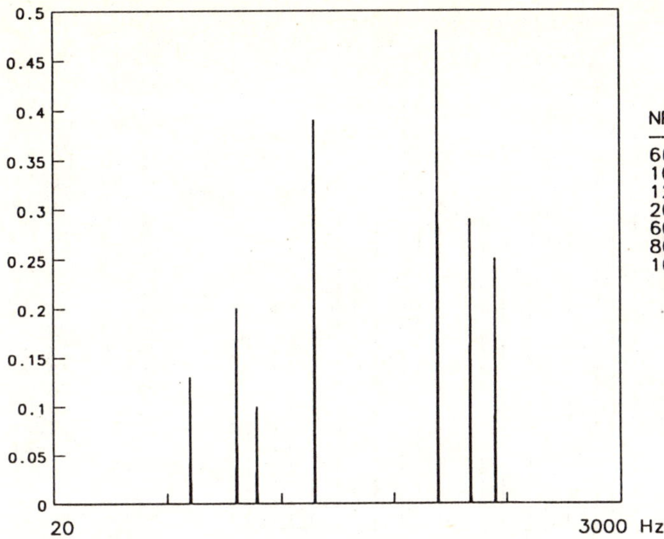

NP	TP	W	T
66.7	62.4	0.13	v
100.1	95.8	0.20	v
120.1	115.7	0.10	v
200.2	195.5	0.39	v
600.6	599.9	0.48	s
800.8	821.5	0.29	s
1001.0	1029.2	0.25	s

Fig. 5.4. Subharmonic-sum spectrum of a complex tone containing the frequencies 600 Hz, 800 Hz, and 1000 Hz, at 60 dB, 55 dB and 50 dB

pitch shift into consideration. The third column (W) comprises the weights of the virtual pitches associated with the frequency and the fourth column shows the type of the pitch (T): "v" means virtual, while "s" means spectrally resolved. Thus, the tones that make up the signal are spectrally resolved. The virtual pitches (above the threshold of 0.10) are located at 66 Hz, 100 Hz, 120 Hz, and 200 Hz NP. The most prominent frequency is 200 Hz. The graph plots only nominal pitches.

Only those pitches that are below a certain frequency (800 Hz or even 500 Hz) are considered to be approximate candidates for virtual pitch. Although tone center recognition relies on the global properties given by the completion image, an algorithm might be used to decide on the most likely pitch by searching for the maximum in the SSHS.

Figure 5.5a shows the spectrum and frequency-amplitude list of the Shepard-chord **DO-MI-SOL**. Figure 5.5b shows the output of TAM as a list of frequency-weight pairs.

For use in the cognition module, the list should be transformed into a format using vectors. Vectors of 36 dimensions have been used in which the frequency range of three octaves (508.31 Hz to 63.54 Hz) is divided into equal intervals. A useful formula to obtain the equal frequency ranges is

$$F_i \;=\; < \; L(H/L)^{i/V}, L(H/L)^{i+1/V} \; > \qquad (5.3)$$

where F_i is the frequency range spanned by the vector element i (ranging from 0 to 35), H is the highest frequency (508.31 Hz), L is the lowest frequency (63.54 Hz) and V is the dimensionality of the vector, which in this case is 36.

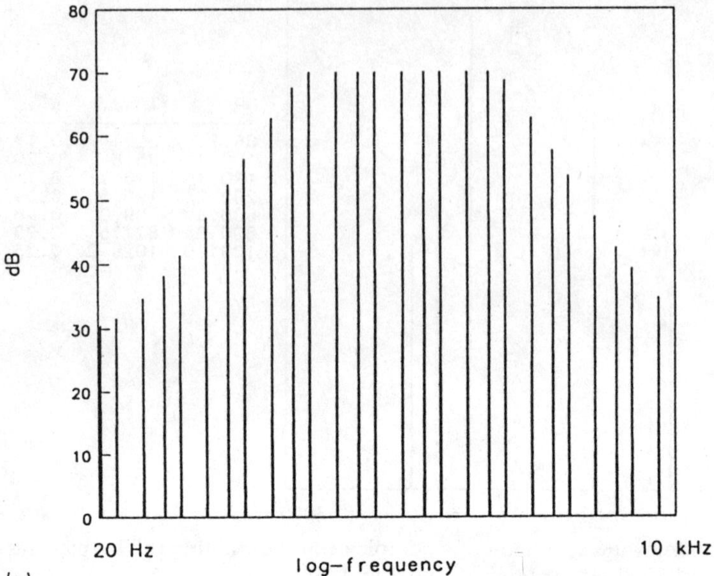

(a)

freq	dB
20.60	30.55
24.50	31.66
32.70	34.77
41.20	38.27
48.99	41.38
65.40	47.27
82.40	52.39
97.99	56.31
130.8	62.73
164.8	67.50
196.0	70.00
261.6	70.00
329.6	70.00
391.9	70.00
523.2	70.00
659.2	70.00
783.9	70.00
1046	70.00
1318	70.00
1568	68.62
2093	62.73
2637	57.61
3135	53.69
4185	47.27
5273	42.50
6271	39.27
8371	34.77

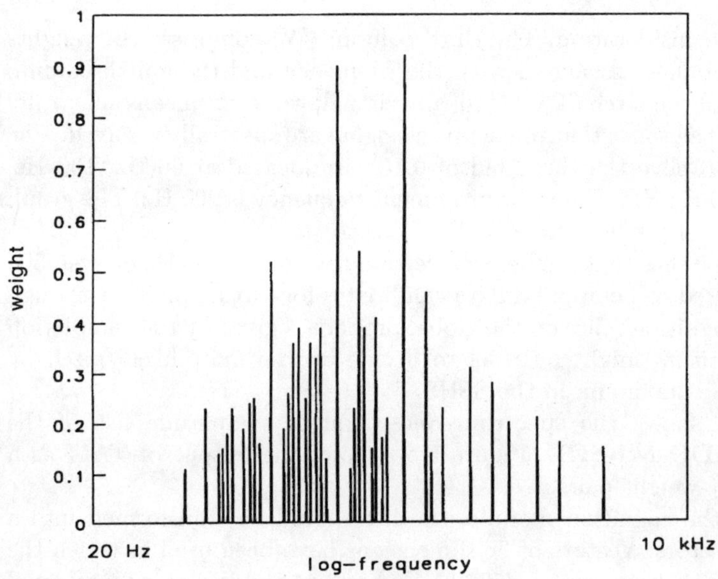

freq	weight
52.3	0.11
65.4	0.23
74.7	0.14
78.4	0.15
82.4	0.18
87.2	0.23
98.0	0.19
104.6	0.24
109.9	0.25
116.2	0.16
130.8	0.52
149.4	0.19
156.8	0.26
164.8	0.36
174.4	0.39
188.3	0.25
196.0	0.36
209.2	0.33
219.7	0.39
232.6	0.13
261.6	0.90
299.0	0.11
313.6	0.23
329.6	0.54
348.7	0.39
376.7	0.15
391.9	0.41
418.6	0.17
439.3	0.21
523.2	0.92
659.2	0.46
697.7	0.14
783.9	0.23
1046	0.31
1318	0.28
1568	0.20
2093	0.21
2637	0.16
4185	0.15

(b)

Fig. 5.5. Spectral representation of the Shepard-chord **DO-MI-SOL**: (a) spectrum and frequency-amplitude pair list, (b) graph and list of virtual and spectral pitches with weight assignments

5.5 VAM: A Place-Time Model

VAM is a place-time model – based on spatial as well as temporal encoding principles of the auditory system. The residue pitch information is based on a periodicity analysis of the auditory nerve images.

Related models have been described by *van Noorden* [5.19], *Assmann* and *Summerfield* [5.1], *Meddis* and *Hewitt* [5.12], and many others. In particular Meddis and Hewitt give a detailed introduction to the power of the pitch identification capabilities of such models, including the perception of ambiguous pitch and pitch shift, and the missing fundamental.

The main difference with the place model is that pitch judgments are not only based on the subharmonic relationships between resolved harmonics, but also on the period of discharge patterns of the auditory nerve fibers that do not resolve the individual partials. The period of these discharge patterns corresponds to the inverse of the missing fundamental. In contrast to Terhardt's model, the subharmonic templates are not supposed to be learned, but have a physiological basis.

The present model is adopted from the computer program by *Van Immerseel* and *Martens* [5.4, 5] but has been modified to satisfy some musical constraints. The synthetic part has been completely redefined.

5.5.1 VAM – The Analytic Part

The analytic part of VAM is based on the different signal processing mechanisms of the auditory periphery. Figure 5.6 gives an overview of the main steps in the process. The first two steps take into account the low- and bandpass filtering of the outer and middle ear (step 1), and the hydro-mechanical bandpass filtering in the cochlea (step 2). The latter is implemented as a bank of asymmetric bandpass filters at distances of one bark (one critical band). Twenty such filters are used in the range of CF 220 Hz to CF 7075 Hz (CF=center frequency). The center frequencies of the filters correspond to the best frequencies of the hair cells (located at one critical band from each other). The auditory nerve fibers associated with the cells are called *channels*. Figure 5.7 shows a compound signal containing frequencies of 600 Hz, 800 Hz, and 1000 Hz at 60 dB, 55 dB, and 50 dB as filtered by a bank of 20 filters. Fig. 4.2 shows the acoustical signal.

The next step involves a transduction from mechanical to neural (step 3). The following features are important:

- **Half-wave Rectification and Dynamic Range Compression**. Due to the polarization effect of the stereocilia, only the positive phase of the signal is captured. The filtered signals are therefore rectified at half-wave. The intensity is coded both by the spike rate of the signal in the neuron and by the activity over different channels. The activity is represented by the probability of firing during a defined time interval. The design by Van

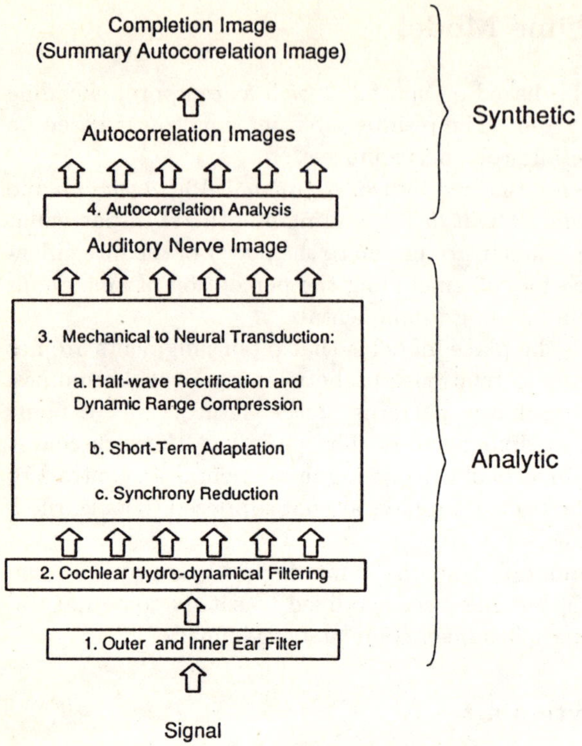

Fig. 5.6. Overview of VAM

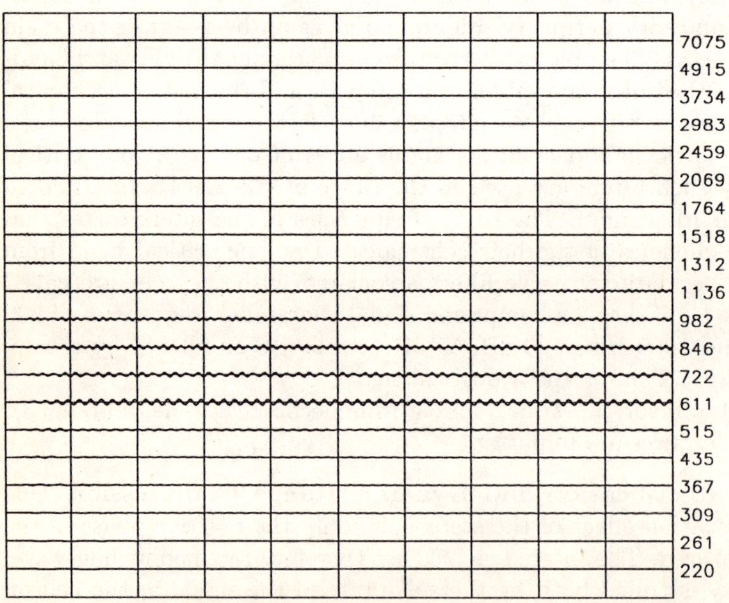

Fig. 5.7. A complex tone containing frequencies of 600 Hz, 800 Hz, and 1000 Hz at 60 dB, 55 dB, and 50 dB is filtered by a bank of 20 filters

Immerseel and Martens is based on a transition zone of maximum 50 dB and the representative values for spontaneous and saturated firing of the neurons are between 0.05 and 0.15 spikes/ms, respectively.

- **Short-term Adaptation**. There is an increased sensitivity after a period of nonstimulation, and a suppressed sensitivity after a period of strong stimulation.
- **Synchrony Reduction**. Because of their inability to synchronize with fast frequencies, neurons transfer temporal envelopes. In the original model this was implemented by a low-pass filtering (250 Hz) of the neural firing patterns, thereby extracting the envelopes of the neural firing pattern. This technique, inspired by the capacity of the auditory nerve patterns to phase-lock to the amplitude modulations in the signal, allows a considerable down-sampling of the original frequency to 500 sa/s. The aim was indeed to develop a real-time model for speech recognition. For musical data, however, 250 Hz might be too low for a fine analysis of periodicity. Therefore, the model has been modified in such a way that the neural firing patterns are filtered at 1250 Hz (sampling rate of 2500 sa/s) in stead of the original 250 Hz. At higher frequencies (smaller distances between the peaks) the latencies of the synchronization stay in a larger range – reflecting the fiber's refractory period. It is known that the limit of synchronization (or phase-locked discharges) of the auditory nerve fibers is about 4000–5000 Hz [5.6]. The phase-locking to integral multiples of the

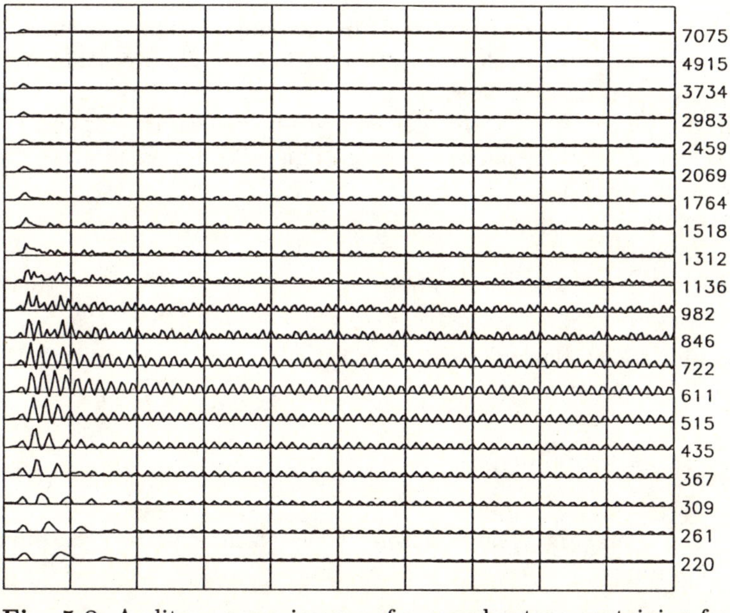

Fig. 5.8. Auditory nerve images of a complex tone containing frequencies of 600 Hz, 800 Hz, and 1000 Hz at 60 dB, 55 dB, and 50 dB

period is lost at about 1100 Hz [5.2]. The current limit is an approximation to the latter range.

The neural firing patterns can be considered signals whose amplitude has been modulated. If two frequencies fall within one critical band, their spectral components are not resolved but they generate a modulated signal in the corresponding auditory channel. See for example channel 9 (CF=846 Hz) in Fig. 5.8. Similarly, since filters overlap, there can be an influence from one channel on the other. Pitch extraction is based on these modulated signals.

Figure 5.8 shows the effect of the mechanical to neural transduction of the signals in Fig. 5.7. Due to dynamic reduction, the former linear representation of amplitude has a pronounced non-linear character. The model operates at a sampling rate of 20000 sa/s but the extraction of the envelope patterns (justified by the synchronization reduction) allows a down-sampling to 2500 sa/s. The auditory nerve image, a vector of 20 elements corresponding to the 20 channels, is therefore updated every 0.4 ms.

5.5.2 VAM – The Synthetic Part

A completion module has been build on top of the peripheral part. Its function is to transform the auditory nerve images into tone completion images. The completion process consists of two steps: a periodicity analysis of the neural firing patterns in each channel, and a sum of the periodicity analyses over all channels.

Autocorrelation. The periodicity analysis in one single channel is implemented by a short-term autocorrelation function (STAF). The resultant image is called an *autocorrelation image.*

To sharpen the peaks in these images, the firing values are clipped to the mean of all values in the analyzed frame. This is common practice in autocorrelation analysis [5.3].

The autocorrelation is defined as

$$R(n) = a(n) \sum_{k=1}^{K-n} s(k)\, s(k + n)\, w(k). \tag{5.4}$$

$R(n)$ is the autocorrelation value at time-lag n (in the range from 1 to K), and $s(k)$ is the signal at k. $w(k)$ takes the form of a decaying exponential as in

$$w(k) = \exp\left(-\frac{k}{T}\right), \tag{5.5}$$

where T is the time constant. The function $a(.)$ is an attenuator according to a parabolic function:

$$a(n) = 1 - \alpha \left(n - \frac{K}{2}\right)^2. \tag{5.6}$$

The parameters are set as follows: $K = 30$ ms, $T = 16$ ms (40 samples) and $\alpha = 0.5$. The frame defined by $K = 30$ ms allows a range of 75 time-lags (between 0.4 ms to 30 ms in steps of 0.4 ms). Using a frame of $K = 30$ ms, the autocorrelation images are computed at intervals of 10 ms.

The attenuator is introduced to decrease the role of those regions in the autocorrelation image whose resolution is either a too dense or too sparse. For example, between 500 Hz and 250 Hz, there are only 6 units to represent the chromatic range, whereas from 250 Hz to 125 Hz, there are 11 such units. Since it must be possible to distinguish semitones, one must be careful that the resolution is fine enough. (For other purposes, interpolation could be used).

Figure 5.9 shows the autocorrelation image of one single 30 ms frame for 20 channels. The points on the ordinate correspond to the time-lags n, from 1 to 75. The abscissa represents the channels, from high frequency (CF 7075 Hz=channel 20) to low frequency (CF 220 Hz=channel 1). The values of the autocorrelation are in proportion to the thickness of the lines. Values below 0.1 % of the highest value have not been plotted.

Summary Autocorrelation. The right part of Fig. 5.9 shows the *summary autocorrelation* or completion image which, for this purpose, is normalized according to the highest value in the vector. The summary autocorrelation is just the sum of the values over all channels.

The value at time-lag 12.5 (here represented by activity at points 12 and 13) corresponds to a period of $12.5 * 0.4 = 5$ ms, that is a frequency of 200 Hz. Without the attenuator, this would be the highest value – its frequency corresponds with the residue pitch of the tone complex. In the present application, however, it is more important to consider the pattern as a whole rather than in terms of exact pitch.

5.5.3 VAM – Examples

Figures 5.10a,b, 5.11 show the output of VAM in a somewhat different way. The signal is a Shepard-sound containing four chords: **CM-FM-Gx7-CM**: the spectrum consists of octave-components within a bell-shaped envelope which favors the region between 500 Hz and 1000 Hz (Sect. 3.3). Each chord has a duration of 500 ms and has a short exponential onset and offset of 30 ms. Between the chords, there is a rest of 200 ms. Figure 5.10a shows the evolution of the *autocorrelation images* in channel 6 (CF=515 Hz). The abscissa shows the time at intervals of 10 ms. The ordinate shows the time-lags of the autocorrelation in one frame (30 ms). In this example, the range corresponds to a residue pitch range of 500 Hz (time-lag 1) to 41.66 Hz (time-lag 56). Due to the fact that the numbering of the time-lags have been shifted by 4, the frequency at a particular time-lag should be calculated as above, using $i + 4$ instead of i. Thus, the formula becomes: $1000/[0.4(i + 4)]$. In this figure, the values below 20 % of the highest value in the sequence (to which all

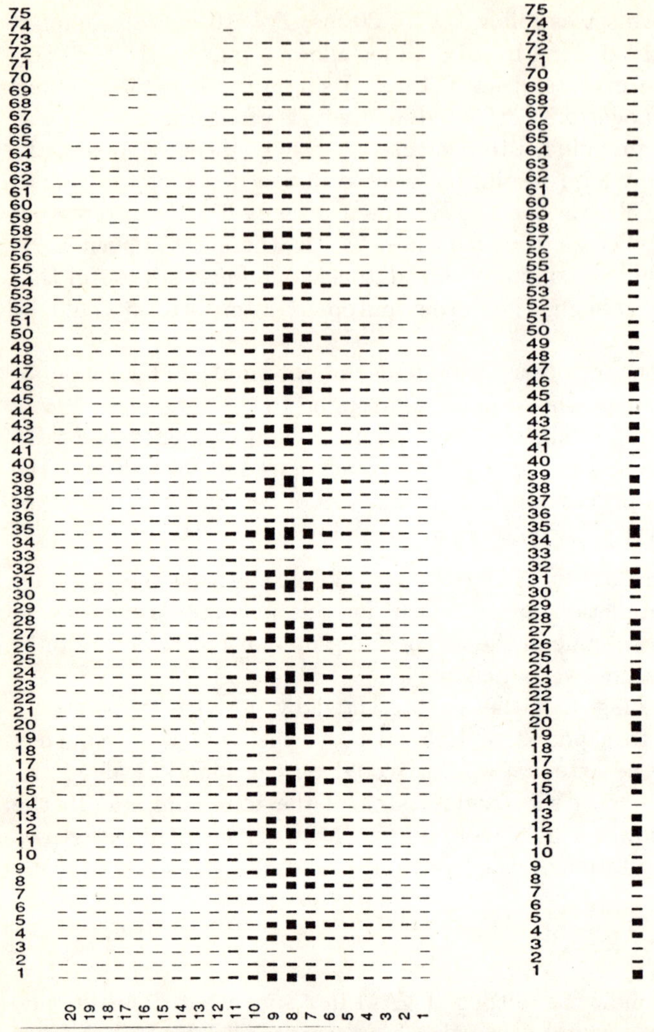

Fig. 5.9. Autocorrelation images in one single 30 ms frame. The left figure shows the autocorrelation images in 20 channels, the right figure shows the summary autocorrelation image

values are normalized) are not represented. Figure 5.10b shows the evolution of the *autocorrelation images* in channel 9 (CF=846 Hz). As in Fig. 5.10a, the values are normalized to the highest value in the sequence.

The *summary autocorrelation images* or completion images result from a sum of the autocorrelation images over all channels. This is shown in Fig. 5.11. The most prominent tone of the first chord is at point 35, its frequency corresponds to: 2500/(35+4)=64.1 Hz (=**DO2**).

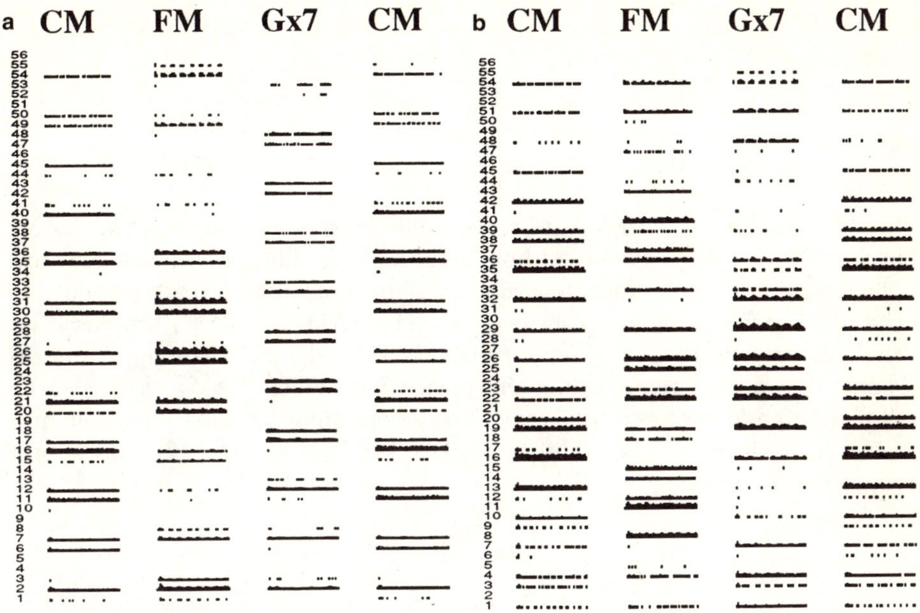

Fig. 5.10. Autocorrelation images in (a) channel 6 (CF=515 Hz) and (b) channel 9 (CF=846 Hz) of the Shepard-chord sequence **CM-FM-Gx7-CM**

Fig. 5.11. Summary autocorrelation images or completion images of the Shepard-chord sequence **CM-FM-Gx7-CM**

As in TAM, images-out-of-time could be obtained by processing the chords independently and calculating a mean completion vector for each chord. With VAM, however, it is more straightforward to work with images as samples of a sound-stream.

5.6 Conclusion

In this chapter, an overview of three auditory models (called SAM, TAM, and VAM) has been given. Although the models differ in concept and implementation, they account for same basic phenomenon in pitch perception: the missing fundamental. The place models (SAM and TAM) are based on the idea that the missing fundamental is related to the common subharmonic of resolved tones, while the place-time model (VAM) also uses the discharge-periods of auditory nerve fibers containing information about unresolved harmonics. The auditory models form the preprocessing parts of a cognition module.

6. Schema and Learning

This chapter gives an introduction to some basic concepts of self-organization, including learning and association. The algorithm of SOM, the self-organizing map, is explained and described.

6.1 Gestalt Perception

For centuries, composers have explored the idea that vertical (or harmonic) writing compels listeners to adopt a listening strategy that favors the fusion of tones, while a polyphonic setting, on the other hand, is more oriented towards the analytic capabilities of the human auditory system. By using transitions from harmonic to polyphonic writing, or ambiguous settings, interesting fluctuations between different listening strategies occur. What is first heard globally, can later be heard analytically and vice versa. Listeners may even try to concentrate on one of these aspects and control perception actively.

The composer G. Ligeti has explored this cognitive dynamics explicitly in many of his compositions. In measure 38 of the *Kammerkonzert*, the whole orchestra plays very softly the same tone in different registers. This leads to extreme fusion but the smallest fluctuation, a musician who suddenly plays a bit louder by accident or voluntarily as indicated on the score, may break down the global listening strategy into an analytical one in which the listener hears distinct instrumental sounds. The change in attention may even be caused by fatiguing effects on the listener's cognitive system.

The musical Gestalt depends on the properties of the stimulus as well as on voluntary aspects. But small fluctuations can have a maximal perceptive impact and may even cause a so-called *catastrophe* of perception (Sect. 3.5).

The change from global pitch to timbre or to individual pitch components suggests a rich dynamics based on different mechanisms. In theories of Gestalt perception [6.6] it is assumed that these mechanisms are in competition with each other.

The cognitive dynamics is not limited to one perceptual category but seems to underlie visual perception as well. Figure 6.1 is a well-known example from Gestalt psychology demonstrating the competition between concurrent stable perception points: the picture of an old woman and a young woman.

Fig. 6.1. Young woman or old woman? [6.5]

The switch from one percept to the other is often due to fatiguing effects or voluntary actions. The stimulus, because of its ambiguity, provides the cues for the transition, although, unlike most examples in music, there is no movement in the stimulus.

Figure 6.2 provides another example of the multi-stability of perception. Three or even more stable percepts are possible in this case. Some people have difficulties in seeing the transparent boxes. A global property of cognitive dynamics seems to be its tendency towards stable points, and distinct factors tend to influence transitions from one state to the other [6.9, 10, 21, 23].

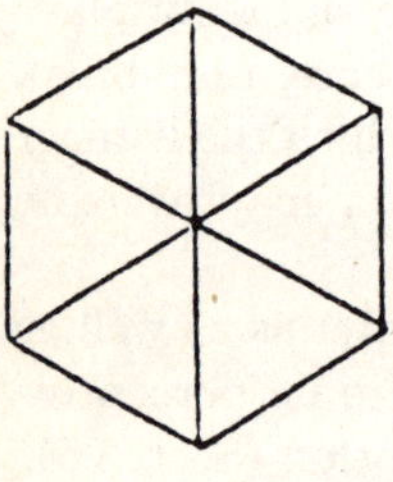

Fig. 6.2. Multistable figure

6.2 Tone Semantics and Self-Organization

A fundamental property of any self-organizing system is that a learned global order can emerge from the local interaction of micro-elements. Such a system is undifferentiated in the beginning but develops co-operative action by interacting with the environment. What kind of global behavior would such a system learn with music as its input? Under which conditions should learning be studied?

Fig. 6.3. Chords, to be played by different instruments

Consider the chords in Fig. 6.3. Assume that these chords are played by a piano, and that the recorded signal – preprocessed with an auditory model – is presented to a learning system. What sort of properties can be expected after training?

When slightly different patterns are presented (for example: the chords are played by violins, instead of a piano) the system should be able to recognize the chord played. There are reasons to believe that a simple associative memory or Hopfield network [6.7] is able to learn the three distinct chords as distinct categories or *fixed points*. Fixed points are local minimum energy states that attract nearby states. But in order for these points to be the only ones, however, the examples should be sufficiently distinct from each other. When they are not sufficiently distinct, interference occurs and *spurious fixed points* emerge. Spurious fixed points are created as an unwanted side-effect of the learning process. In most applications one tries to avoid these effects [6.8]. Sometimes, however, spurious attractors may be quite useful. *Serra* and *Zanarini* [6.20] point out that spurious attractors can be considered as the expressions of self-organization – the system's *autonomous interpretation* of input.

Do spurious attractors lead to semantics? At this point it is necessary to consider a more elaborate set of data. Assume that instead of three chords, a representative set of tonal chords is given. The set contains major triads, minor triads, major sevenths, minor sevenths, augmented sevenths, and so on. Hundreds of different chords could be played by different instruments in different settings (inversions). Interference of all these chords is probably unavoidable. Will the system be able to recognize the chords as individual chords, independent of the timbre of the instrument? Will the system be able to recognize the chord type, rather than the specificities of the chord inversions? Will interference ultimately lead to the perception of tonality? Will the system be able to take into account contextual semantics?

To study these complexities in an ordered way, a distinction has been made between two forms of self-organization:

- **Self-Organization as Learning**. The model used in this book is SOM, the *Self-Organizing Map* (also known as the Kohonen-map [6.11, 12]). SOM

will be discussed in this chapter and applied in order to study the emergence of a schema for tone perception in Chaps. 7–8.
- **Self-Organization as Association**. The model is called TCAD, which stands for *Tone Center Attractor Dynamics* [6.14-16]. TCAD will be discussed and evaluated in Chaps. 9–10.

Both are complementary and it is a good methodological strategy to separate them from the outset.

6.2.1 Self-Organization as Learning

To be able to describe tone semantics in terms of attractor dynamics it is necessary to consider first the way in which a schema and an associated set of attractors, come into existence. One plausible answer is that they emerge from the interaction with the musical environment. As such, global tonal knowledge can be considered as the output of self-organization processes applied on input patterns.

Neurobiologists, such as *Changeux* [6.1] and *Edelman* [6.2, 3] describe this process as the adaptation of the organism to the environment by the formation of perceptual categories and internal taxonomies on which the responses to the environment are based. The brain develops responses and adapts itself to the stimuli of the environment by the refinement of structure and function. Adaptation occurs in the refinement of the neural structure early in life and by the refinement of synaptic functions. During learning, the sensorial information is the determining factor. Classification and organization lead to reduced representations of relevant objects in the environment.

6.2.2 Self-Organization as Association

In the last decade, brain research and the theory of self-organization has developed into a vast framework for the study of intelligence. The approaches include: natural selection mechanisms [6.2], autopoiesis and self-steering [6.18], reaction-diffusion systems [6.22], synergetics, and general dynamic systems that operate close to points of instability [6.6]. Applications are found in astrophysics, biology, chemistry, and ethology...

In music, the idea that the cognitive basis of tone semantics could be expressed through the concepts of stability, attraction, and tendency or movement have been circulating for a long time. In the first half of the 19th Century, *Fétis* [6.4] already attempted to describe the tonic in terms of an attraction dynamics and he was followed by many other musicologists – albeit at a purely metaphorical level.

A first step towards an operational account involves the definition of stable points, that is: points that would attract other nearby perception points. SOM is used to show that such points can emerge through processes of learning. But once established, they may attract the unstable perception points of the information flow that is continuously processed by the auditory system. Fluctuations in the tone pattern, may thus cause transitions from one tone center to another. In the tonal system, thc change from one tonality to another is known as a *modulation.*

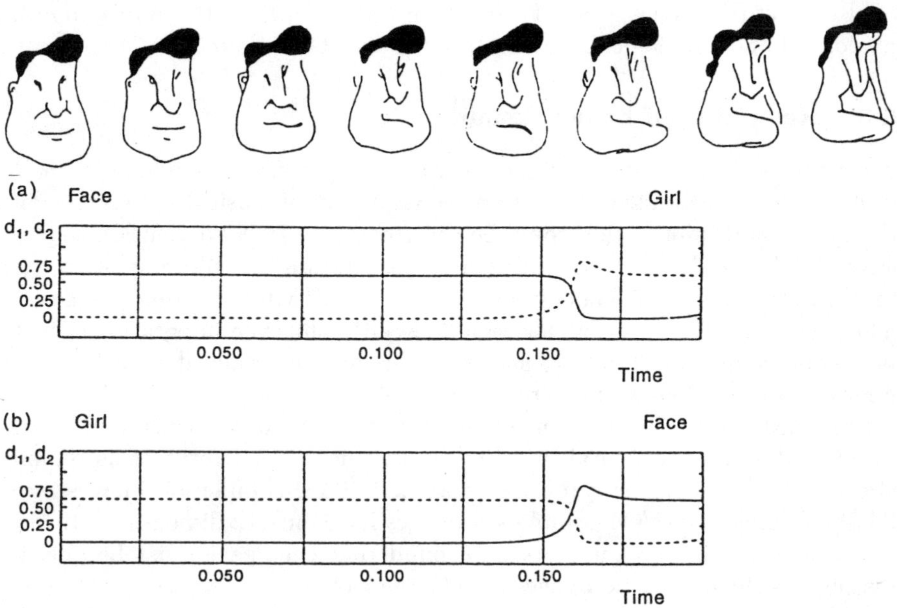

Fig. 6.4a,b. Hysteresis in visual perception. The graph shows the delay in going from face to woman (and vice versa) [6.5]

Modulation is often characterized by an effect of *hysteresis* [6.10, 21]: a retardation in the transition from one point to the other because of attraction, much like what happens in Fig. 6.4. From left to right, the stable percept of a face determines the interpretation of the ambiguous figures. From right to left, the stable percept of a woman sitting determines the interpretation of the ambiguous figures. The two interpretations do not intersect at the same point. The overlapping results from the retardation effect that favours stable perception points. In music perception, like in visual perception, the effect of hysteresis is often cancelled by an interpretation process in which the perceived information of a recent past is reconsidered in the light of new evidence.

6.3 SOM: The Self-Organizing Map

In the subsequent chapters, learning is modelled by computational principles adopted from Kohonen's self-organizing map [6.11, 12]. SOM is perhaps the best-known algorithm for data-driven self-organized learning. It is based on principles of cortical information processing, in particular the formation of feature maps as representations of real-world objects in the memory. The neurobiological evidence for the existence of such maps has been discussed in Sect. 4.2.3.

Relevant properties of SOM are dimensional reduction, the ability to form analogical and topological representations, and its statistical aspects.

6.3.1 Reduction of Dimensionality

The objects of the surrounding world can be represented as points in an N-dimensional embedding space. From the viewpoint of musical signals, the auditory spectral information could be described in a 15980-dimensional space (from 20 Hz to 16 kHz). Building a representation on the inner hair cells in a row between the helicotrema and the oval window would suggest a 3400-dimensional space (one for each hair cell), while the integration of just-noticeable frequency variations suggests there are 640 adjacent steps and thus a 640-dimensional space for pitch-objects.

In VAM (Sect. 5.5) this range is reduced to only 20 hair cells, but a temporal analysis of the neural firing patterns generates patterns of 56 dimensions. Other dimensions can be used as well. TAM-completion images have 36 dimensions, but SAM-completion images have only 12 dimensions. In this discussion it is important to bear in mind that the space must be able to distinguish different tone images as different points.

The most important dimensionality reduction then occurs at the cognitive level. According to *Kohonen* [Ref. 6.11, p. 125], the tendency to compress information forming reduced representations of the most relevant facts, without loss of knowledge about their interrelationships (the so-called *second-order isomorphism*), is a fundamental principle of intelligent information processing.

A basic property of the cognition module is that it reduces the dimensionality of the input to the hidden dimensions of the implicit regularity. Although the images themselves are high-dimensional, the mutual relationships between the images can be low-dimensional. For robot applications, the high-dimensional embedding space is typically reduced to a *three*-dimensional geometry [6.19]. In tone center perception we assume that the low-dimensional space is *two*-dimensional.

6.3.2 Analogical and Topological Representations

Reduction of dimensionality is biologically interesting because a reduced representation is much more compact and easier to store and to work with than the multi-dimensional original. The reduced representation should eliminate most of the unnecessary redundancy and noise from the input, while still containing its most relevant aspects.

SOM produces a mapping from an N-dimensional space onto a two-dimensional one, which is represented by a grid of neuron-like elements. For a given 56-dimensional completion image, a two-dimensional "bubble" of units in the grid will be activated with one unit in the center of this bubble coming on most strongly. Thus what is stored in memory is not the original image but some of its relational aspects mapped onto a two-dimensional map.

The neurons in the grid are organized in such a way that similar input patterns (according to some metric) turn on nearby bubbles of activity in the two-dimensional grid. Then, by inspecting the layout of the grid after the network has been trained – that is, seeing which regions are activated by which input patterns – a particular reduced dimensionality and topological organization of the input data can be discovered, similar to the spatial organizations found by multi-dimensional scaling. The similarity relationships in the observable world are represented in the memory representation and the metric is reflected in topological relationships. Such a map is both *analogical* and *topological* (Sect. 2.4).

6.3.3 Statistical Modeling

A third important characteristic of the Kohonen model concerns the distribution of input patterns. When patterns are represented by points in a N-dimensional embedding space, it may happen that the points are not equally distributed over the space. The self-organizing map will reflect this distribution on the map. The spatial resolution will tend to follow the non-uniformity of the probability distribution of the input patterns and mirror the amplifications – in a way which is similar to the differently amplified topographic maps of the cerebral cortex [6.19].

6.4 Architecture

The self-organizing map can be thought of as comprising two layers of neuron units: the two-dimensional grid layer where the action occurs and an input layer of neurons that relays information from the perception module (the auditory model). Between these two layers are a set of synapses that form full interconnections; that is, every grid neuron has a set of synapses coming into it which connect it to every input neuron.

Figure 6.5 shows the basic structure of a single grid neuron (input lines, synapses, and output lines are shown), while Fig. 6.6 shows the self-organizing map's two-dimensional grid of these neurons, each linked via synaptic connections to the neurons in the input layer. These connections are shown for only two neurons in the network, but, as we mentioned above, in the computer model all connections are present.

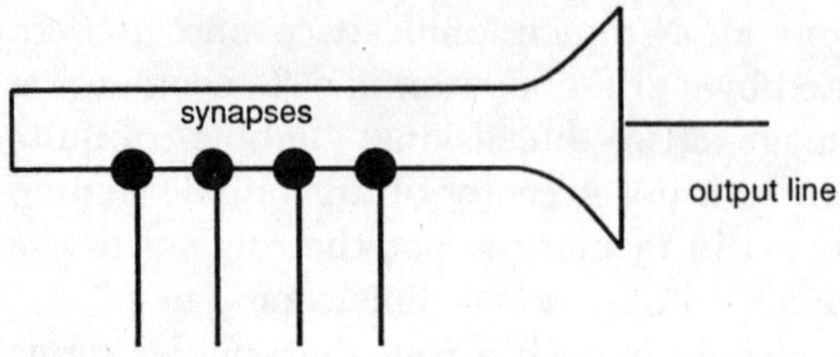

Fig. 6.5. Basic structure of a neuron showing input lines, synapses and output line

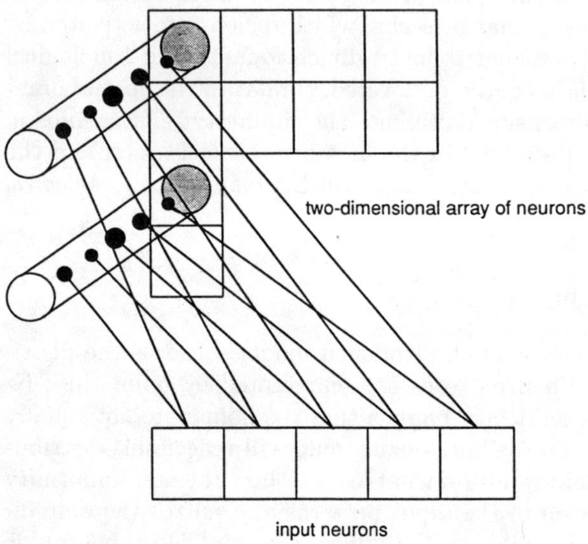

Fig. 6.6. Basic architecture of the self-organizing map, showing an array of input neurons and a two-dimensional array of grid neurons. The input neurons are connected to the synapses of the grid neurons

6.5 Dynamics

Neurons are computational units: they accumulate activity that comes from other neurons and produce on this basis the activity for the output to other neurons. With real neurons, the activity is all-or-none: a neuron integrates signals at the synapses and fires if the membrane difference potential goes beyond the threshold. High activation is thus translated into fast firing.

In the model, the activity of the neurons is represented by the firing probability over the time interval at which the system is updated – just like in VAM. Unlike VAM, however, the time interval is not critical for the algorithm because SOM works "out-of-time" (the training patterns are chosen in random order).

The activity from an input neuron to a grid neuron is then modulated by a connection strength called *synaptic efficacy.* The synaptic efficacies are weighting the inputs to the grid neurons. Like in real neurons, the weighted sum must exceed a threshold or bias before the grid neuron can be activated. But since SOM has only one layer of synapses, the activation of a neuron can be computed as a weighted sum of the activations of the neurons in the input layer. The activation of the grid neuron i is simply:

$$act_i = \sum_j a_j * w_{ji}. \tag{6.1}$$

In this equation, a_j is the activation of input neuron j and w_{ji} is the synaptic efficacy (weight) of the link from input neuron j to grid neuron i. In other words, act_i is the correlation of the activation vector with the vector of the synaptic efficacy (the weight vector).[1]

Network training starts with no particular organization among the grid neurons: all synapses start with random weights. But as the map is trained by repeated presentations of the input data set, it begins to self-organize, and topological patterns of certain regions of grid neurons responding to certain sorts of inputs begin to appear.

Self-organization consists of making small adjustments to the synaptic strengths over and over again in response to a sort of competition among the grid neurons, until the strengths perform a stable dimensionality reduction of the input data and yield the appropriate grid organization. The connection strengths can be regarded as the long-term memory of a system that adapts itself to the environment.

The adaptation of the synaptic weights proceeds as follows:

1. Initialize the synaptic weights by giving them a random value.
2. Choose a *learning rate*, which governs the rate at which the synaptic strengths are adjusted, and a *neighborhood radius*, which will be used to specify the size of the bubble of activity among the grid neurons in response to a particular input pattern.
3. Select an input pattern at random, and use it in the following steps.
4. Present the input pattern to the network, and find the grid neuron which becomes most active in response to this input. This is done by taking the grid neuron whose synaptic efficacy vector has the least distance to the input vector according to the Euclidian metric:

$$d_j = \sqrt{\sum_i (a_i - w_{ij})^2}. \tag{6.2}$$

5. For all grid neurons that fall in the neighborhood radius of the selected neuron, adapt the synapses leading to the neurons according to the formula:

$$w(t+1) = w(t) + \alpha(t) * (a(t) - w(t)), \tag{6.3}$$

where $w(t+1)$ is the synaptic efficacy vector at time t+1, $w(t)$ is the synaptic vector at time t, $\alpha(t)$ is the learning rate at time t, and $a(t)$

[1] The correlation is only one measure of similarity and alternative measures can be used. One of these is the Euclidian distance, which will be used in the model as well. See Appendix C for more explanation.

the input vector at time t. This learning rule will tend to make the winning neuron and its neighbors more likely to win the competition for this particular input pattern, and those like it, in the future.
6. If necessary, decrease the learning rate and neighborhood radius.
7. Go back to step (3) for each input pattern of the cycle, and repeat for subsequent cycles.

Summing up, for each presentation of an input pattern, find the best-matching grid neuron, and increase the match at this neuron and its topological neighbors. In this way, the bubbles of activity in response to particular input patterns are formed, and nearby bubbles will respond to similar patterns.

6.6 Implementation

SOM has a straightforward parallel implementation and for that reason its implementation has been realized on a computing system with an arbitrary number of processors, using EXPRESS [6.13, 17]. An efficient simulation depends on a balanced equilibrium between the size of the network and dimension of the input vectors on the one hand, and the amount of processors on the other. Most of the simulations for this book were realized on a PC-based Transputer system with 4 T800 processors having each 2 MB of internal memory. The processor topology is efficient for relatively small networks (400 neurons and 12-dimensional input vectors) and small data sets: profile analysis shows a gain factor of about 3.8 when 4 processors are used instead of one. In order to be able to process a large amount of data, the program has recently been ported to a nCUBE/2 located at CNUCE-CNR (Pisa, Italy).[2]

6.7 Conclusion

Self-organization is an essential part of the way in which organisms form an internal representation of the environment. Two types of self-organization have been distinguished: self-organization as a learning process and self-organization as an associative process. In this chapter, the focus has been on a popular model called SOM. SOM unfolds a dimensional reduction of the perceived objects in an analogical and topological representation. Although SOM is static (no associative dynamics is involved), the classes obtained by learning from examples can be interpreted as stable points. The relevance for an associative dynamics will be discussed in Chap. 9.

[2] A simulation based on the model VAMSOM (Sect. 8.6) has been carried out with the nCUBE/2 using 8 (16 MB) custom processors each with 2.3 Mflops/s peak performance. The simulation, nevertheless, took about 20 hours, which illustrates that SOM is computational intensive.

7. Learning Images-out-of-Time

This chapter evaluates the data-driven long-term learning of a self-organizing artificial neural network. The training data are tone images-out-of-time. This means that the images are not temporally related to each other so that a straightforward reduction of data is possible.

The aim is to show that:

1. The responses emerging from learned tone images are structured.
2. The structure is compatible with the mental representation of tone centers.
3. The results are robust, and can be obtained with different auditory models.

7.1 SAMSOM

This section evaluates SAMSOM – a model consisting of the auditory model SAM and the self-organizing model SOM. Previous results with a similar model have been reported in *Leman* [7.5–7]. The computer results are characterized by the selection and preprocessing of training data, the system or network parameters, the evolution of learning and aspects of ordering.

7.1.1 Selection of Data

The training set consists of 115 different chord images. Each chord appeals to an auditory object-out-of-time.

A chord is built up with minor and major-third intervals. A major triad, consisting of a major-third interval [M] and a minor-third interval [m], has the interval structure [M,m]. If the root of the chord is given in addition to the interval structure, then the notes of the chord can easily be reconstructed. For example, given the root **DO**, the notes of the DO-major triad chord are obtained by taking **DO** as a basis, adding first the major third [M] (which gives **MI**), and then the minor third [m] (which gives **SOL**). The shorthand notation for the DO-major triad, comprising the notes **DO-MI-SOL**, is **CM**. To make a clear distinction between notes, chords, and tone centers, the following notation is used:

– *notes* are labeled by **DO**, **DO♯**, **RE**, ... etc.
– *chords* are labeled by their root **C**, **C♯**, **D**, ... etc to which the chord specification for major, minor etc... is appended into one string, such as in **CM**, **Cm**, **Co** ..., **C♯M**, **C♯m**, **C♯o** ..., **DM**, **Dm**, **Do** ....
– Finally, the *tone centers* are labeled as **C**, **C♯**, **D**, ... for the *major* tone centers, and **c**, **c♯**, **d**, ... for the *minor* tone centers.

The chords used in the training set of this study are listed in Table 7.1. Since SAM relies on reduced Shepard-tones, inversion and octave spreading is not taken into account.

Table 7.1. Overview of Chords

Number	Nature	Structure	Notation
12	major triads	[M,m]	**CM**
12	minor triads	[m,M]	**Cm**
12	diminished triads	[m,m]	**Co**
4	augmented triads	[M,M]	**C+**
12	major seventh chords	[M,m,M]	**CM7**
12	minor seventh chords	[m,M,m]	**Cm7**
12	dominant seventh chords	[M,m,m]	**Cx7**
12	half diminished seventh chords	[m,m,M]	**CO7**
12	augmented seventh chords	[M,M,m]	**C+7**
12	minor with major seventh chords	[m,M,M]	**Cm-7**
3	diminished seventh chords	[m,m,m]	**Co7**

The pattern of the augmented triads and the diminished seventh chords is repeated after **E♭+** and **Do7**. That is, the notes of the next augmented triad, **E+** (**MI-LA♭-DO**), are the same as in **C+**; and the notes of the diminished seventh chord **E♭o7** (**MI♭-FA♯-LA-DO**), are the same as in **Co7**. Therefore, the number of possible chords for these two classes is reduced.

In SAMSOM, the images are based on simple spectral patterns. The S-patterns of the chords based on **DO** are:

1 0 0 0 1 0 0 1 0 0 0 0 **CM**
1 0 0 1 0 0 0 1 0 0 0 0 **Cm**
1 0 0 1 0 0 1 0 0 0 0 0 **Co**
1 0 0 0 1 0 0 0 1 0 0 0 **C+**
1 0 0 0 1 0 0 1 0 0 0 1 **CM7**
1 0 0 1 0 0 0 1 0 0 1 0 **Cm7**
1 0 0 0 1 0 0 1 0 0 1 0 **Cx7**
1 0 0 1 0 0 1 0 0 0 1 0 **C07**
1 0 0 0 1 0 0 0 1 0 0 1 **C+7**
1 0 0 1 0 0 0 1 0 0 0 1 **Cm-7**
1 0 0 1 0 0 1 0 0 1 0 0 **Co7**

7.1.2 Preprocessing

The completion images are obtained by transforming the S-patterns into R-patterns by (5.1). The above patterns thus result in:

1.83	0.10	0.45	0.33	1.10	0.70	0.25	1.00	0.33	0.85	0.20	0.00	**CM**
1.60	0.20	0.25	1.33	0.10	0.95	0.00	1.00	0.83	0.35	0.20	0.33	**Cm**
1.10	0.20	0.58	1.10	0.20	0.75	1.00	0.00	1.08	0.10	0.20	0.83	**Co**
1.33	0.60	0.45	0.00	1.33	0.60	0.45	0.00	1.33	0.60	0.45	0.00	**C+**
1.83	0.35	0.45	0.33	1.60	0.70	0.25	1.33	0.43	1.05	0.20	1.00	**CM7**
1.85	0.20	0.25	1.83	0.10	0.95	0.33	1.10	1.03	0.35	1.20	0.33	**Cm7**
2.08	0.10	0.45	0.83	1.10	0.70	0.58	1.10	0.53	0.85	1.20	0.00	**Cx7**
1.35	0.20	0.58	1.60	0.20	0.75	1.33	0.10	1.28	0.10	1.20	0.83	**C07**
1.33	0.85	0.45	0.00	1.83	0.60	0.45	0.33	1.43	0.80	0.45	1.00	**C+7**
1.60	0.45	0.25	1.33	0.60	0.95	0.00	1.33	0.93	0.55	0.20	1.33	**Cm-7**
1.10	0.20	1.08	1.10	0.20	1.08	1.10	0.20	1.08	1.10	0.20	1.08	**Co7**

Since normalization is not done in SOM, it should be done before the data enter the cognition module. In this example, the completion images are normalized by (C.6) (Appendix C). This gives the set:

0.51	0.03	0.12	0.09	0.30	0.19	0.07	0.28	0.09	0.24	0.06	0.00	**CM**
0.47	0.06	0.07	0.39	0.03	0.28	0.00	0.30	0.25	0.10	0.06	0.10	**Cm**
0.39	0.07	0.21	0.39	0.07	0.27	0.36	0.00	0.39	0.04	0.07	0.30	**Co**
0.43	0.19	0.15	0.00	0.43	0.19	0.15	0.00	0.43	0.19	0.15	0.00	**C+**
0.44	0.08	0.11	0.08	0.38	0.17	0.06	0.32	0.10	0.25	0.05	0.24	**CM7**
0.44	0.05	0.06	0.44	0.02	0.23	0.08	0.26	0.25	0.08	0.29	0.08	**Cm7**
0.47	0.02	0.10	0.19	0.25	0.16	0.13	0.25	0.12	0.19	0.27	0.00	**Cx7**
0.35	0.05	0.15	0.41	0.05	0.19	0.34	0.03	0.33	0.03	0.31	0.21	**C07**
0.32	0.20	0.11	0.00	0.44	0.14	0.11	0.08	0.34	0.19	0.11	0.24	**C+7**
0.41	0.12	0.06	0.34	0.15	0.24	0.00	0.34	0.24	0.14	0.05	0.34	**Cm-7**
0.34	0.06	0.33	0.34	0.06	0.33	0.34	0.06	0.33	0.34	0.06	0.33	**Co7**

The complete list of images consists of 115 such labeled chords. They are obtained by rotating the **C**-chords over all notes of the chromatic scale.

7.1.3 Network Specifications

A discussion of the network specifications entails three aspects: the network architecture, the network size, and the network training.

- **Network Architecture**. The *architecture* is a two-dimensional neural grid with a torus structure: the upper and lower sides of the network, as well as the left and right sides, are connected to each other.
- **Network Size**. Since the interest is in both global and local responses to chords, a network of 20 by 20 neurons has been chosen. The reader may be surprised by the relatively large size of the network. There are indeed 400 neurons for only 115 different training data! But the reason for this

large network, compared to the small amount of training data, is that there must be enough space in the network for each chord to be represented by a different neuron. This condition is necessary for obtaining an idea of how the data get topologically related to each other. Recall that the chords are *prototypes* or class-objects, standing for chords in a particular inversion, octave setting, and timbre. The network size assures enough place to separate these chords and fully display its analogical and topological properties. The real power of the neural network becomes evident when a large amount of training data are used.
- **Network Training**. The network dynamics has been described in Sect. 6.5. In the present simulation, the training patterns appear in random order during each training cycle. The *learning rate* is set to 0.02 and this value is kept constant during the training session. The *neighborhood radius* is set to 18 and decreases after every 10 cycles so that after 180 cycles the radius will be 0. The program stops after 300 training cycles.

7.1.4 Aspects of Learning

Figure 7.1 shows the error evolution of the training session. The training cycles, during which the whole data set is run through in random order, are represented on the horizontal axis. The vertical axis represents the mean Euclidian error, that is: the sum of the distances of each pattern to the neuron with the largest response divided by the number of patterns. Characteristics of the error evolution are:

- The error approaches zero after 300 cycles, which implies that the patterns have an almost exact representation in the network. This result is reasonable because, compared to the size of the training set (115 patterns), the size of the network (400 neurons) is such that individual neurons can become tuned to individual patterns.
- The error decreases with decreasing neighborhood. During adaptation, a large neighborhood means that the neurons become tuned to a mean pattern (which differs from all the rest). During this phase, a global organization can take place but the mean error does not decrease significantly. With a smaller neighborhood, the neurons become adapted to the individual patterns and its close relatives. During this phase, a local organization can take place. After 100 cycles the mean error decreases rapidly, because the neurons become more and more specific. A demarcation between global and local organization typically occurs when the radius is smaller than half of the network size.

Figure 7.2 show the network response to the completion image of **CM** (the C-major triad) after 1, 60, 150, and 300 cycles respectively. Each block represents the activation (Euclidian distance) of a neuron in the grid, with the size corresponding to the amount of activation. The resulting values are normalized by rescaling the highest and lowest values in the interval of 1 to 0.

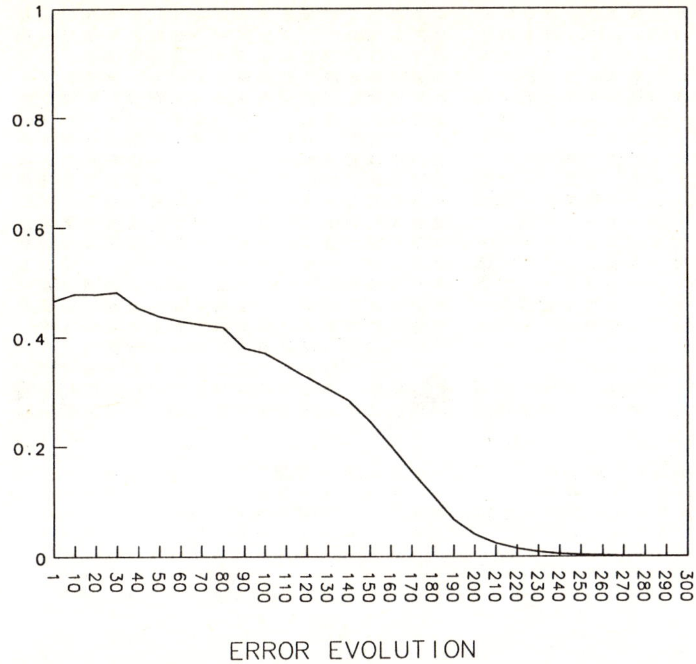

Fig. 7.1. SAMSOM error evolution over 300 cycles

After the first training cycle (Fig. 7.2a), the memory is still totally chaotic (as it was initially) and there is no clear center of response to the input. After some training, however, the network responds by activating a group, or "bubble", of neurons instead of one single neuron (Fig. 7.2b). The bubbles emerge because of the fact that the neurons in the neighborhood of the highest responding neuron are adapted to the input.

The activation of neurons in response to **X** is called the *Response Region of* **X**. The short hand notation is RR(**X**). Each RR(**X**) has a neuron which is responding best to **X**. This is called the *Characteristic Neuron of* **X** or CN(**X**).

Both the notion of *response region* and *characteristic neuron* are important for an understanding of what is going on during the process of self-organization. The response regions (RRs) evolve in such a way that the activation evolves towards a smaller and more clearly defined bubble. Also, the center of the bubble, characterized by the characteristic neurons (CNs), changes during the learning process. For example, after 1 cycle (Fig. 7.2a), the CN(**CM**) is located at point [1,6].[1] After 60 cycles (Fig. 7.2b), the CN moved to point [20,3], and after 150 cycles (Fig. 7.2c) to point [9,1]. After 300 cycles (Fig. 7.2d) it is localized at point [8,1].

---

[1] The coordinates should first be read horizontally, reading from left to right, and then vertically, reading from bottom to top.

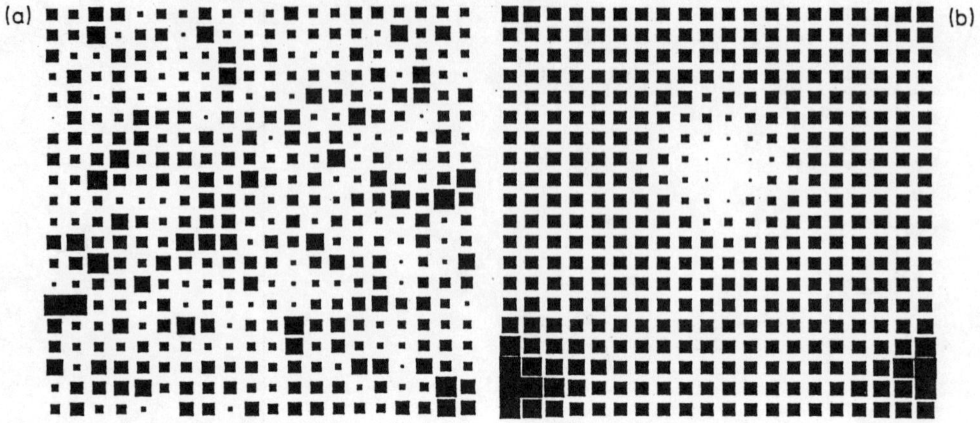

Fig. 7.2. SAMSOM network responses to the chord **CM**: (a) network response after 1 learning cycle, (b) network response after 60 learning cycles, (c) network response after 150 learning cycles, (d) network response after 300 learning cycles. Each block represents the activation of a neuron in the grid, with the size corresponding to the amount of activation

A general overview of all CNs is obtained by presenting all the patterns to the network and labeling each CN with the corresponding input label.

Examples are shown in Fig. 7.3, for 1, 60, 150 and 300 training cycles, respectively. One can observe the *migration* and *clustering* of CNs on the map during learning:

– Migration means that the CNs may change their location on the map. Without migration the network would never be self-organizing.

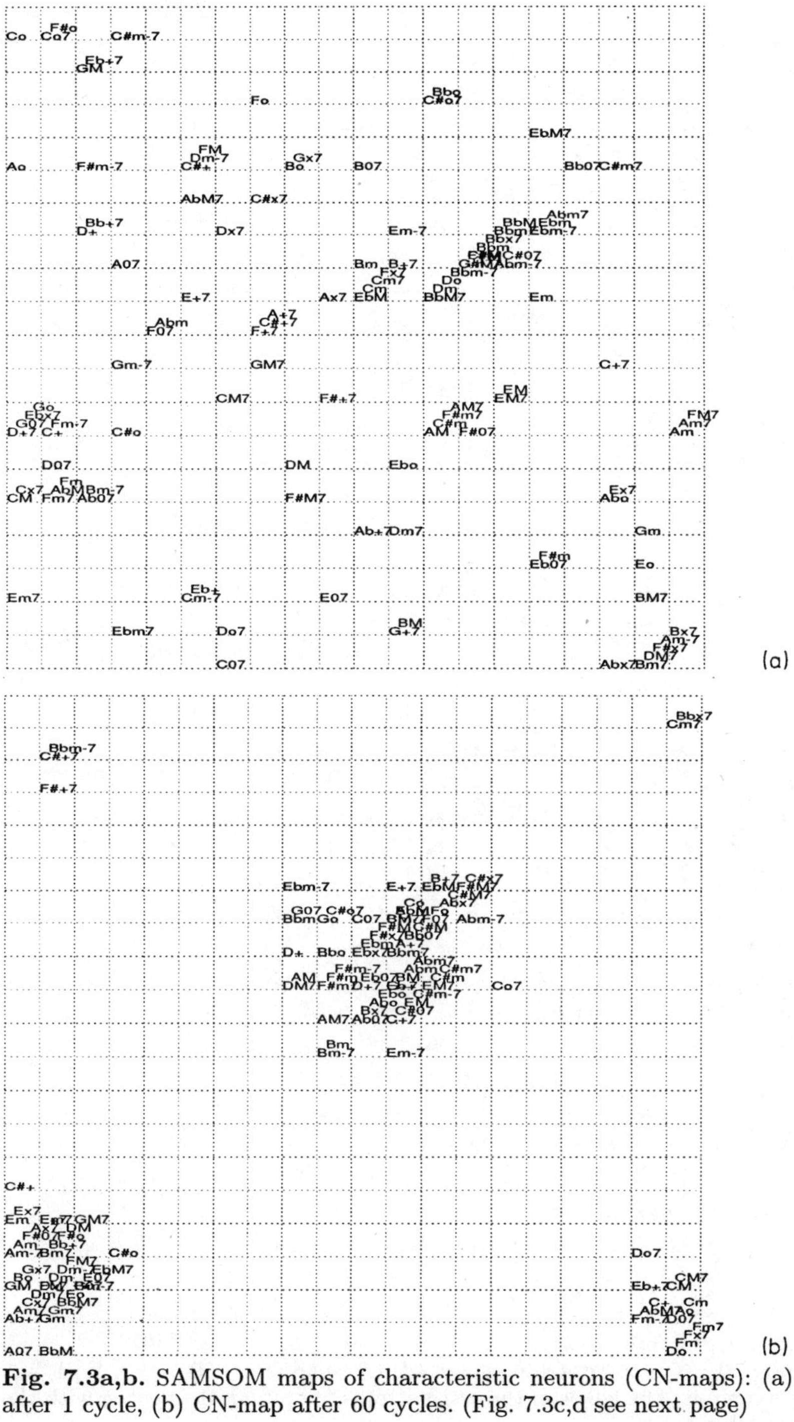

Fig. 7.3a,b. SAMSOM maps of characteristic neurons (CN-maps): (a) CN-map after 1 cycle, (b) CN-map after 60 cycles. (Fig. 7.3c,d see next page)

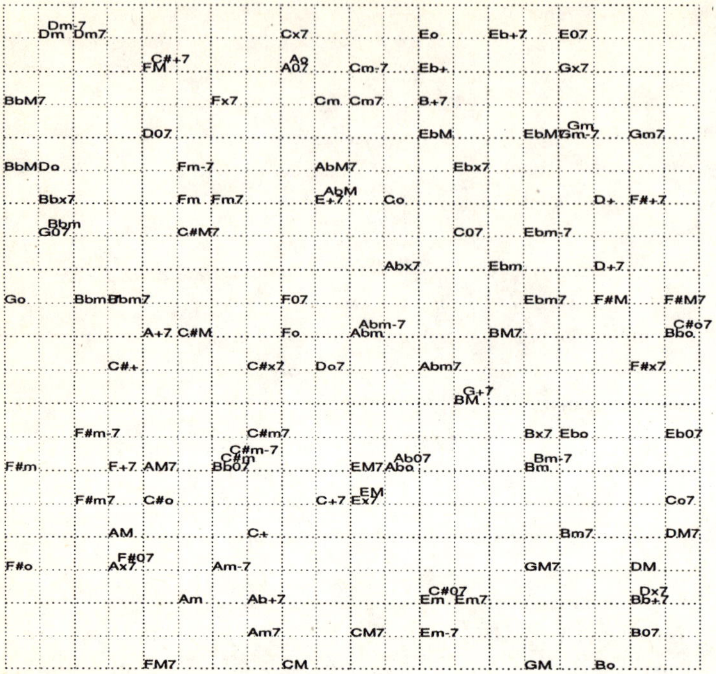

(c)

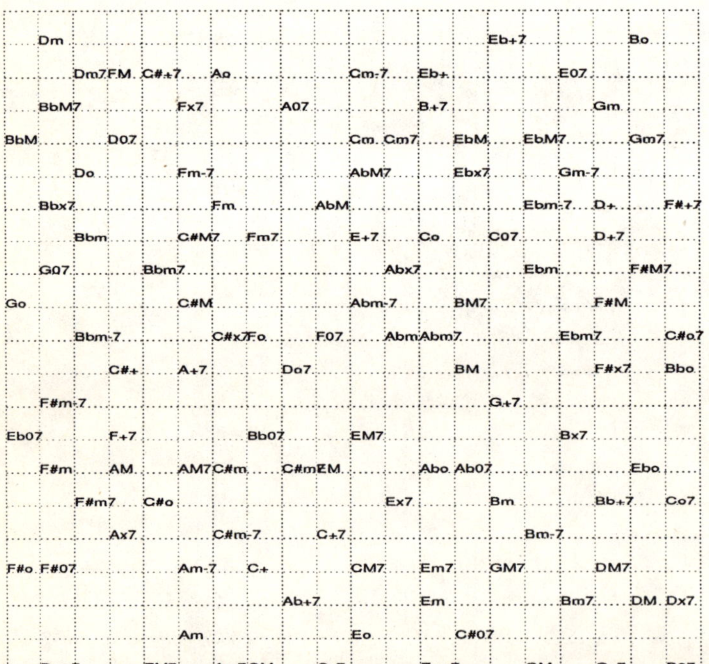

(d)

Fig. 7.3c,d. (c) CN-map after 150 cycles, (d) CN-map after 300 cycles. Each box stands for a neuron. The labels of the test patterns are put on the neuron with the highest activation

– Clustering means that one single neuron might be characteristic for several inputs. If this is the case, the global error cannot be zero.[2]

The response is chaotic in the beginning but starts taking shape in an early stage. Figure 7.3b is typical for the early categorization process. The neighborhood radius of 12 neurons is still large, so that a local specification of neurons is not yet possible. The result is a grouping of the chords into two categories with typical clusters. After 100 cycles (not shown here) 4 CN-groups can be observed. When the neighborhood radius is further reduced, the CNs can migrate toward other places so that clustering is resolved (Fig. 7.3).

A change in the neighborhood radius, learning rate and rate of neighborhood decrease has an effect on learning and ordering. Experiments have shown that learning is faster when the neighborhood is decreased after every 3 learning cycles or when the learning rate is higher (e.g., 0.1). Most of these changes have an effect on the ordering as well but there is a certain variability of parameters so that the observations about global and local ordering, migration and clustering of CNs hold for different parameter settings. This leads to the conclusion that the learning of a particular data set in the self-organizing map is robust [7.6].

7.1.5 Ordering and Emergence

An analysis of the emergent structure of a schema should take into account the different image categories, in particular the relationship between chords images and tone center images.

Chord Relationships. Chord relationships provide information about the interchangeability of chords. It is possible to distinguish two types of ordering: local and global ordering. Local ordering is related to a property which is called the interchangeability of chords. Global ordering is related to the circle of fifths.

Local Ordering and Chord Interchangeability. Some aspects of local ordering can be deduced from Fig. 7.3d (after 300 cycles). Consider CN(**CM**) at point [8,1] and its nearest characteristic neuron CN(**Am7**)at point [7,1]. Both responses are related to each other. One may assume that this holds because the notes in **CM** (**DO-MI-SOL**) form a subset of those in **Am7** (**LA-DO-MI-SOL**). As a consequence, the completion image will be closely related and this relationship will be reflected on the map.

Looking carefully, one observes that the minor-seventh [m7] chords are all in the vicinity of the associated major [M] chords. The major-seventh [M7] chords are also close to the major [M] chords (**E♭M** and **E♭M7**, or **A♭M** and **A♭M7**) and the minor-seventh [m7] chords are close to the minor [m] chords.

[2] Also, if the global error is not zero, clustering occurs. But, of course, this depends on the amount of training patterns. If there are more distinct training patterns than neurons (which is normally the case), clustering will always occur.

In general, related chords are represented close to each other on the map. Several techniques can be used to explore the ordering in more detail:

- **Limited and Characteristic Response Regions.** A first technique is based on the concept of *limited response region* or LRR. A LRR takes into account only those neurons whose activation is above a certain threshold. It is thereby possible to restrict the LRR to CNs only.
  Assume that N is the set of all neurons in the network, and C the set of all CNs. In the present setup, C is a subset of N. The RR generated by a particular **X** is a function that maps N onto a domain of activations (from 0 to 1):
  $$\mathrm{RR}(\mathbf{X}) : \mathrm{N} \to [0,1]. \tag{7.1}$$
  Restricting N to all CNs reduces the RR to only those neurons that are characteristic. This set is called the *characteristic response region of* **X** or CRR(**X**). Expression (7.2) says that if C is a subset of N, then CRR(**X**) is a subset of RR(**X**).
  $$\textit{if}\, \mathrm{C} \subset \mathrm{N}\ \textit{then}\ \mathrm{CRR}(\mathbf{X}) \subset \mathrm{RR}(\mathbf{X}). \tag{7.2}$$
  The set can be further restricted by considering only those CNs whose activation is above a certain threshold h:
  $$\mathrm{CRR}(\mathbf{X})^h \subset \mathrm{CRR}(\mathbf{X}). \tag{7.3}$$
  Below, the CRRs contain the labels of neurons whose similarity with the input is above $h = 0.4$. The CRRs are generated by the images **CM**, **FM**, and **GM**. The values between brackets represent the Euclidian distances (scaled for all units of the network between 1 and 0):

  CRR(**CM**)$^{0.4}$ =
  { **CM**(1.00), **Cx7**(0.71), **Am7**(0.71), **CM7**(0.69), **A♭+7**(0.69), **Am**(0.55), **FM7**(0.51), **Eo**(0.49), **E07**(0.48), **Am-7**(0.47), **A07**(0.47), **Em7**(0.46), **Em**(0.46), **Cm**(0.46), **Ax7**(0.44), **Cm-7**(0.43), **C+7**(0.43), **F♯07**(0.42), **C+**(0.42), **Fm-7**(0.41)}

  CRR(**FM**)$^{0.4}$ =
  { **FM**(1.00), **Fx7**(0.71), **Dm7**(0.71), **FM7**(0.69), **C♯+7**(0.69), **Dm**(0.55), **B♭M7**(0.51), **Ao**(0.49), **A07**(0.49), **Dm-7**(0.47), **D07**(0.47), **Fm**(0.46), **Am7**(0.46), **Am**(0.46), **Dx7**(0.44), **Fm-7**(0.43), **F+7**(0.43), **C♯+**(0.42), **B07**(0.42), **B♭m-7**(0.41)}

  CRR(**GM**)$^{0.4}$ =
  { **GM**(1.00), **Gx7**(0.71), **Em7**(0.71), **GM7**(0.69), **E♭+7**(0.69), **Em**(0.55), **CM7**(0.51), **Bo**(0.49), **B07**(0.49), **Em-7**(0.47), **E07**(0.47), **Gm**(0.46), **Bm7**(0.46), **Bm**(0.46), **Ex7**(0.44), **Gm-7**(0.43), **G+7**(0.43), **E♭+**(0.42), **C♯07**(0.42), **Cm-7**(0.41)}

An invariant pattern can be extracted from these lists. Using the degrees (**I**, **I♯**, **II**, ..., **VII♭**, **VII**) to characterize a chord, the CRR for $h = 0.5$ can be written as follows:

CRR(**M**)$^{0.5}$ =
{ **IM**(1.00), **Ix7**(0.71), **VIm7**(0.71), **IM7**(0.69), **VIb+7**(0.69), **VIm**(0.55), **IVM7**(0.51) }

Similarities between images can be obtained by more direct methods – not based on neural networks. By calculating the correlation coefficient of the tone image of **CM** with all the other images, one obtains a list with values between 1 and −1. Below, such a list is shown for the 20 highest correlation values:

CM(1.00), **Am7**(0.88), **Cx7**(0.85), **CM7**(0.85), **A♭+7**(0.84), **Am**(0.69), **Eo**(0.64), **FM7**(0.58), **E07**(0.57), **A07**(0.55), **Em**(0.54), **Cm**(0.54), **Em7**(0.52), **Am-7**(0.52), **C+**(0.51), **C♯o**(0.45), **F♯07**(0.44), **Cm-7**(0.44), **Fm-7**(0.41), **Ax7**(0.39).

As expected, this list is similar to the CRR(**CM**)$^{0.4}$ and the conclusion is that the network is able to capture the relationships among the images. The network, however, has a straightforward topological representation of the image relationships and provides a carrier for the representation of structured information with inherent generalizing capabilities – including emergent relationships between emergent images. When the network is tested with new patterns, and even more in simulations with huge amounts of data, the power of the network will become evident.
The analysis of the relationships between the network responses can also be based on the *intersection* of CRRs. The intersection of limited CRRs may reveal typical features, such as in:

$$\mathrm{CRR}(\mathbf{CM})^{0.5} \cap \mathrm{CRR}(\mathbf{FM})^{0.5} = \{\mathbf{FM7}\}$$

$$\mathrm{CRR}(\mathbf{CM})^{0.5} \cap \mathrm{CRR}(\mathbf{GM})^{0.5} = \{\mathbf{CM7}\}.$$

It reveals the fact that major seventh chords (because of the seventh) have some similarity with the chord of the dominant. A problem with this technique, however, is that the calculation of the intersection of CRRs depends too much on the threshold. The lower the threshold, the larger the border and the more chords that appear in the intersection.

- **Overlap of Response Regions**. A second technique for the analysis of network responses – one that is independent from the concept of CN – is based on the overlap of RRs. Each RR can indeed be considered as a vector of 400 dimensions and the overlap can be accounted for in terms of a similarity measure. An example is given below. The comparison of RR(**CM**) with all other RRs produces a list which is related to the above results. Again, only the 20 highest values are given. The calculations are based on the correlation coefficients:

CM(1.00), **Am7**(0.91), **Cx7**(0.88), **CM7**(0.87), **A♭+7**(0.86), **Am**(0.75), **Eo**(0.70), **A07**(0.67), **FM7**(0.65), **Am-7**(0.63), **E07**(0.61), **Em**(0.56), **Cm**(0.55), **Em7**(0.54), **C+**(0.52), **Cm-7**(0.50), **F♯07**(0.46), **C♯o**(0.46), **Ax7**(0.46), **Ao**(0.45).

To summarize, aspects of local ordering and chord interchangeability may be deduced from at least two observational methods:

– the calculation of CRR for a particular image,
– the calculation of the similarities between the RRs of all images with respect to the RR of a particular image.

The second method is more reliable because two images **H** and **G** may generate the same CNs on the map, while their RR can be different. In other words, if CN(**H**) = CN(**G**) then it does not necessarily follow that RR(**H**) = RR(**G**)[3]. In general, however, the map of CNs will give a initial idea of the relationships between the represented objects and can be of help as a first aid to a structural analysis. Finer observations should be based on an analysis of the relationships between RRs.

The musical interpretation of the local organization of the chords is interesting. The lists indeed suggest possibilities for chord *substitution*. For example, in the chord-sequence

CM-FM-GM-CM

one may replace FM, GM, and even CM by a related chord, such as in:

CM-Dm7-Gx7-CM

or

Am7-Fx7-Em7-CM

By transposition, the correlations for other chords can easily be deduced from the above lists.

But a remark about the interchangeability of chords is at its place here. The notion of interchangeability is applied to images-out-of-time, and when put in a sequence, one should consider the temporal interactions. Temporal interactions are not taken into account in this chapter and for that reason the above cadences can only prove interchangeability from a static point of view. Interchangeability does not explain the creation of tonal tension or relaxation.

[3] The definition of CN could be given in terms of the centroid of the activation pattern, rather than just the neuron with the highest activation. In the present study, however, this alternative has not been taken into account.

Global Organization and the Circle of Fifths. Apart from a local organization, a global organization can be observed by considering the overlap of the RRs of particular chord-types. For example, the correlation of RR(**CM**) with the RRs(**M**) (all RRs of the major triad chords) produces the list:

CM(1.00), **C♯M**(-0.46), **DM**(-0.02), **E♭M**(0.09), **EM**(0.05), **FM**(0.43), **F♯M**(-0.64), **GM**(0.40), **A♭M**(0.05), **AM**(0.12), **B♭M**(-0.05), **BM**(-0.44).

RR(**FM**) and RR(**GM**) resemble most those of RR(**CM**), while RR(**FM**) and RR(**GM**) are much less similar when compared to each other. This becomes obvious when comparing Fig. 7.4 (containing RR(**FM**) and RR(**GM**)) to Fig. 7.2d (containing RR(**CM**)).

Fig. 7.4. SAMSOM network responses to the chords **FM** and **GM**: (a) network response to **FM**, (b) network response to **GM**

On Fig. 7.3d, the global structure can be revealed by drawing a line between a particular CN and its two most similar CNs of the same chord-type. For major triads, the obtained structure will reflect the circle of fifths.[4]

The global structure can be further revealed by drawing lines for minor triads ([m]-chords). As for the major triads, the connections should be made according to the similarity relationships between the RRs. For RR(**Cm**), the most similar RRs are generated by **Fm** and **Gm**:

---

[4] Under ideal circumstances, the most similar CNs are closests on the map. In network architectures that do not have a torus structure, or where the training patterns produce more deviations, this principle is not very reliable because deformations of the representation may occur. For that reason, too, it is better to rely on method 2 (comparison of RRs), rather than on method 1 (calculation of the CRR).

Cm(1.00), C♯m(-0.46), Dm(-0.06), E♭m(0.07), Em(0.03), Fm(0.40), F♯m(-0.64), Gm(0.38), A♭m(0.03), Am(0.10), B♭m(-0.06), Bm(-0.47).

It is straightforward to compare this structure with the map of tone profiles (keys) obtained by multi-dimensional scaling of psychological data (Fig. 2.7). In this figure, the major triads are organized as one single circle of fifths, with the minor triads between the relative and parallel major.

Tone Center Relationships. So far, however, the network was tested with the training patterns. The question is whether the learned neuronal functions in a self-organized system can produce structured responses to patterns that were not used in the training. This section evaluates the response to tone images that are supposed to represent more general tendencies in pitch perception – the *tone center images.*

There are three possible ways to proceed: the first approach is based on music theory and practice, the second is based on psychological data, and the third is based on classification techniques.

1. **Tone Centers by Pattern Integration**. Music theory and practice provide evidence that cadences generate tone stability. In Western music, tone centers are characterized by such cadences. Since cadences are sequences of chords it should be possible to compose a pattern that expresses tone center.

   A simple method to generate such a tone center image is by taking the mean of the chord images that occur in typical cadences. This method is called *simple pattern integration.* This technique has been used by *Parncutt* [7.9], who observed a high similarity of the resulting patterns with psychological data. The chord-degrees and types that are selected for the major key and the minor key are:

	IM-IVM-VM-IM
Major tone center:	**IM-IIm-VM-IM**
	IM VIm VM IM
	Im-IVm-VM-Im
Minor tone center:	**Im-IIo-VM-Im**
	Im VI♭M VM Im

   The tone center image for **C** and **c** is computed by taking the mean of super-imposed chord images :

   $$\mathbf{C} = \frac{\mathbf{6CM + Dm + FM + 3GM + Am}}{12}$$

   and

   $$\mathbf{c} = \frac{\mathbf{6Cm + Do + Fm + 3GM + A\flat M}}{12}.$$

   The resulting images for **C** and **c** are:

```
0.36 0.05 0.21 0.08 0.24 0.21 0.05 0.31 0.07 0.24 0.09 0.10 C
0.34 0.11 0.15 0.25 0.11 0.25 0.02 0.31 0.24 0.09 0.12 0.14 c
```

The other tone center images are obtained by rotation.

2. **Tone Centers from Psychological Data**. An alternative method for obtaining representations of tone center images is based on the tone profiles in the work of Krumhansl (Sect. 2.4). This method is interesting because the tone profiles provide data which are independent from the data used for training the neural network. Below, the Krumhansl tone profiles for **C** and **c** have been normalized according to Expression (C.6) (Appendix C):

```
0.39 0.14 0.21 0.14 0.27 0.25 0.15 0.32 0.15 0.22 0.14 0.18 C
0.19 0.38 0.16 0.21 0.32 0.15 0.21 0.15 0.28 0.24 0.16 0.20 c♯
```

All other tone profiles can be obtained by rotation. There is a high similarity between the tone profiles and the integrated images of the previous paragraph. The correlation coefficient of the integrated image of **C** with the tone profile of **C** is 0.96. The correlation coefficient of the integrated image of **c** with the tone profile of **c** is 0.89.
The high correlation suggests that the results obtained by simple integration will give similar results as when tone profiles are used. Unfortunately, the tone profiles are always 12-dimensional and cannot be used in simulation experiments that rely on a higher dimension. In such cases, the integrated image can be used as a test pattern. Their supposed reliability, however, is based on the high similarity in the 12-dim case.

3. **Vector Quantization**. A third approach is based on the idea that the self-organizing map is a pattern classifier that groups neurons into classes. Candidates for such a decision process are nearest neighbor and vector quantization methods [7.1, 3]. The basic idea of vector quantization is to divide the network into a number of regions, where each region is represented by a pattern. With any pattern as input, it is represented by the pattern of the corresponding region. When the objective is to minimize the distortion of the region vectors by learning, the technique is called *learned vector quantization*. Vector quantization can be used for the determination of tone center regions, given that the network has been trained for a particular piece of music.

The tone center relationships reveal the emergent properties of the network in response to integrated images or tone profiles. The relationships give a general idea of the underlying *response structure* of the schema.

In what follows, the integrated images are used as test patterns for SAMSOM. The response structure is obtained by comparing the RRs with each other. Afterwards, this response structure is compared with the structure in the psychological data. All comparisons are based on the computation of the correlation coefficient.

Figure 7.5 shows a combined map of RR and CNs for **C**, **F** and **G**, respectively. In Fig. 7.5a, the major and minor tone centers are shown and connected according to the procedure outlined in Sect. 7.1.5. In Fig. 7.5, only the RR is different. The overlap is apparent and will be reflected in the response structure.

The response structure is shown in Fig. 7.6. The dotted curve shows the relationships between the RRs of the tone centers about **C** and **c**, respectively. The full curve displays the relationships between the (Krumhansl) psychological data about tone profiles (Fig. 2.8). A measure of correspondence between the results of our model and the psychological data is obtained by comparing the full curve and the dotted curve. The curves of Fig. 7.6a yield a correlation coefficient of 0.97, those of Fig. 7.6b a correlation coefficient of 0.98. This shows that the network response structure to tone center images resembles the analogical structures found in psychology.

The result of presenting the (Krumhansl) tone profiles as test patterns to the network is shown in Fig. 7.7. The response structure to **C** and **c** is represented by the dotted curves, while the correlations between the tone profiles are represented by full curves. The similarity between dotted and full curve is 0.99 for major centers (Fig. 7.7a) and 0.99 for minor centers (Fig. 7.7b). We therefore conclude that tone center relationships in the network have a close similarity with the schemata for pitch perception.

Tone Center/Chord Relationships. A tone center image is based on a sequence of chords and therefore it implies a context – although a rather abstract one since it is here considered as an image-out-of-time. Nevertheless, the mutual relationship between tone centers and chords may reveal aspects of a context-sensitive semantics – one in which the context is limited to very typical chord progressions (cadences). The relationships give an idea of stabilizing or destabilizing effects in a given context.

Below, the example is based on tone profile representations of tone center images and a restricted set of [M]-, [m]-, and [o]-chords. The relationship between RR(**C**) and the chord-RRs is given by the following list:

CM(0.94), **C♯M**(-0.51), **DM**(0.15), **E♭M**(0.08), **EM**(0.07), **FM**(0.44), **F♯M**(-0.65), **GM**(0.61), **A♭M**(-0.06), **AM**(0.12), **B♭M**(0.02), **BM**(-0.35), **Cm**(0.51), **C♯m**(-0.25), **Dm**(0.33), **E♭m**(-0.53), **Em**(0.66), **Fm**(0.05), **F♯m**(-0.34), **Gm**(0.43), **A♭m**(-0.33), **Am**(0.71), **B♭m**(-0.48), **Bm**(0.01), **Co**(-0.16), **C♯o**(0.45), **Do**(-0.05), **E♭o**(-0.30), **Eo**(0.68), **Fo**(-0.30), **F♯o**(0.31), **Go**(-0.13), **A♭o**(0.03), **Ao**(0.43), **B♭o**(-0.31), **Bo**(0.27).

The chord images that best fit with the tone center image of **C** are **CM**, **Am** and **Eo**. The list can be compared with the psychological data of *Krumhansl* [Ref. 7.4, p. 171–172] and *Bruhn* [7.2] but the results are not conclusive for all chord types. Except for the [M]- and [m]-chord images,

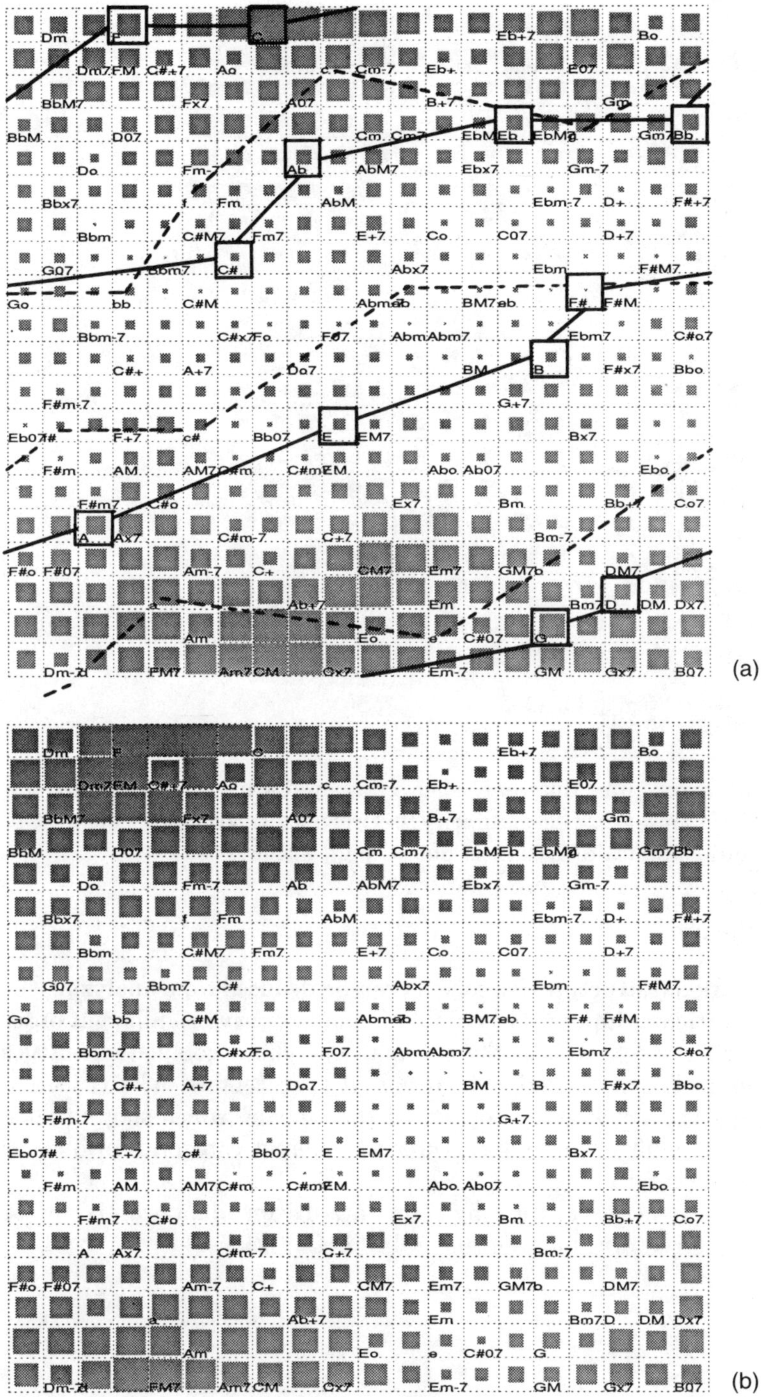

Fig. 7.5a,b. SAMSOM maps for network response and characteristic neurons (RR/CN-map) to tone centers: (a) RR/CN-map for **C**, (b) RR/CN-map for **F** (Fig. 7.5c see next page)

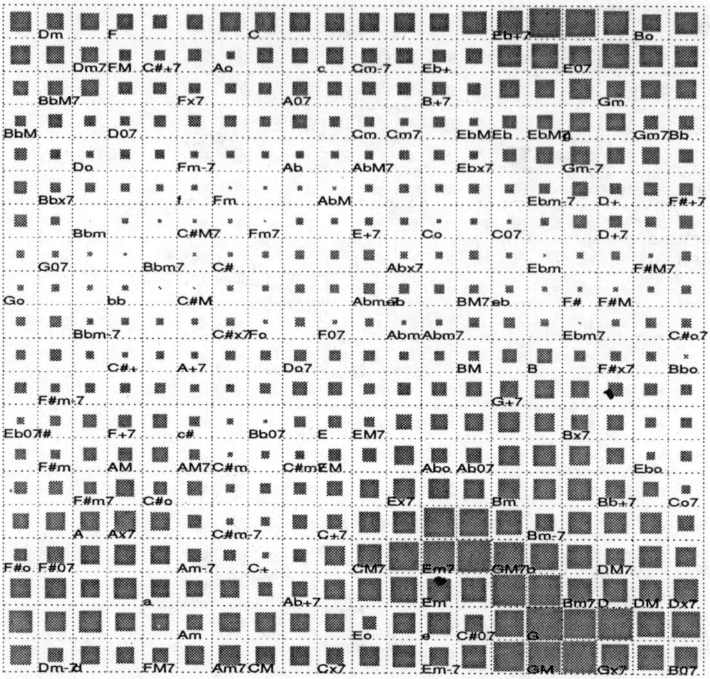

(c)

Fig. 7.5c. (c) RR/CN-map for **G**

the results are even quite different when compared with Bruhn's data. On the other hand, however, the similarity between the psychological data of Krumhansl and the data of Bruhn is not so high either.

7.1.6 Conclusion

The representational structure of a schema can be deduced from analyzing the responses to particular inputs. The completion images have an inherent order and this ordering is reflected in the schema. The chord completion images underlay a more general ordering in terms of tone centers. The reflection of this ordering is an emergent property of the schema, which shows that a schema is a carrier for response structures to different categories of images: chords and tone centers.

The response structure of the long-term data-driven adaptation is similar to the response structures found in mental representations.

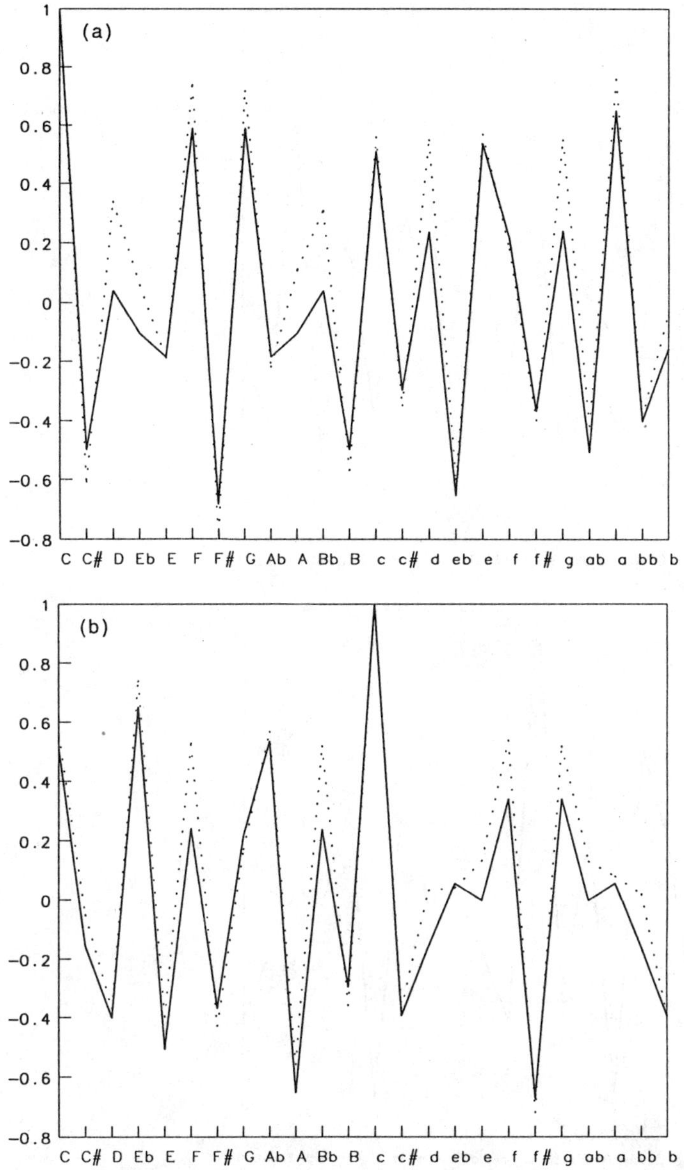

Fig. 7.6. SAMSOM network response structures of tone center images: (a) network response structures to the tone center image of **C**, (b) network response structures to the tone center image of **c**. The structures show the similarity of RR(**C**) and RR(**c**) with respect to all other RRs of tone center images

7.2 TAMSOM

This section evaluates TAMSOM – a model which combines the auditory model TAM and the self-organizing model SOM. The changes to the network specifications illustrate the robustness of the model.

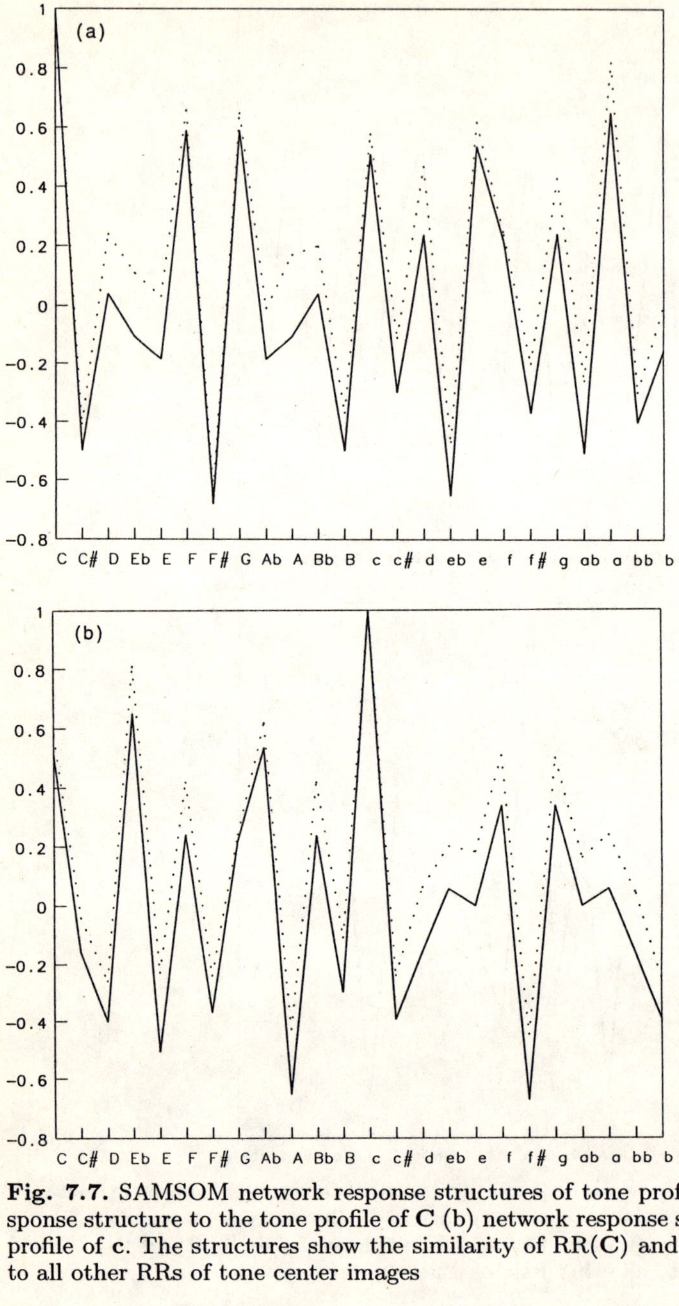

Fig. 7.7. SAMSOM network response structures of tone profiles: (a) network response structure to the tone profile of **C** (b) network response structure to the tone profile of **c**. The structures show the similarity of RR(**C**) and RR(**c**) with respect to all other RRs of tone center images

7.2.1 Selection of Data and Preprocessing

The training set contains the 115 different images-out-of-time described in Sect. 7.1.1. The completion images are obtained by the preprocessing stages discussed in Sect. 5.4.3:

1. Compute a list of frequency-amplitude (Hz-dB) pairs for each chord based on the S-patterns.
2. Process the list into a list of virtual and spectral pitches.
3. Select the frequency-amplitude pairs for the frequencies between 63.54 Hz and 508.31 Hz. This range covers 3 octaves and is considered to be representative for the residue pitches.
4. Transform the list of frequency-amplitude pairs into completion images (vectors) with 36 dimensions. Each element of a vector covers a frequency range of equal log-frequency steps.
5. Normalize each pattern according to the Euclidian norm.

7.2.2 Network Specifications

The network architecture has a torus structure as in SAMSOM. The embedded dimensionality of the completion images is 36. The learning rate is 0.02, the radius of the neighborhood is 18 and the radius decreases after every 10 cycles. The program is stopped after 500 learning cycles and the learning rate is decreased linearly so that after 1000 cycles, the rate becomes zero. (In SAMSOM this value was kept constant.) The output activation is computed as the inner product of the synaptic vector and the input vector. (Euclidian distance was used in SAMSOM). The small difference in parameter settings illustrate the robustness of the model.

7.2.3 Ordering and Emergence

The error evolution of the learning process approximates zero after 500 cycles and displays similar characteristics as observed in SAMSOM (Fig. 7.1).

Figure 7.8 shows the map of CNs. There is a clustering at the points [15,19], [20,17], [7,16], [11,12], [2,9], [18,9], [7,7], [8,4], and [13,1]. At these points, chords with many common tones – (mostly) related to each other by a minor or major third – join together. The effect is probably due to the TAM preprocessor and it might be necessary either to extend the frequency range of the residue pitches to 800 Hz instead of 508 Hz, or to take a finer resolution in the vector.

Chord Relationships. The local and global organization of chords can be analysed as in SAMSOM but a detailed analysis is left to the reader.

Tone Center Relationships. Aspects of emergent structures are investigated by analyzing the response of the network to patterns that stand for tone center images. In this application, however, the tone profiles cannot be used because TAM-images have a 36-dimensional space. The testing patterns are therefore synthesized with the help of the integration technique. Note also that the structure of the patterns does not allow the use of the rotation technique in SAMSOM (Sect. 7.1.2). Each tone center image must therefore be computed separately by integrating the appropriate chords.

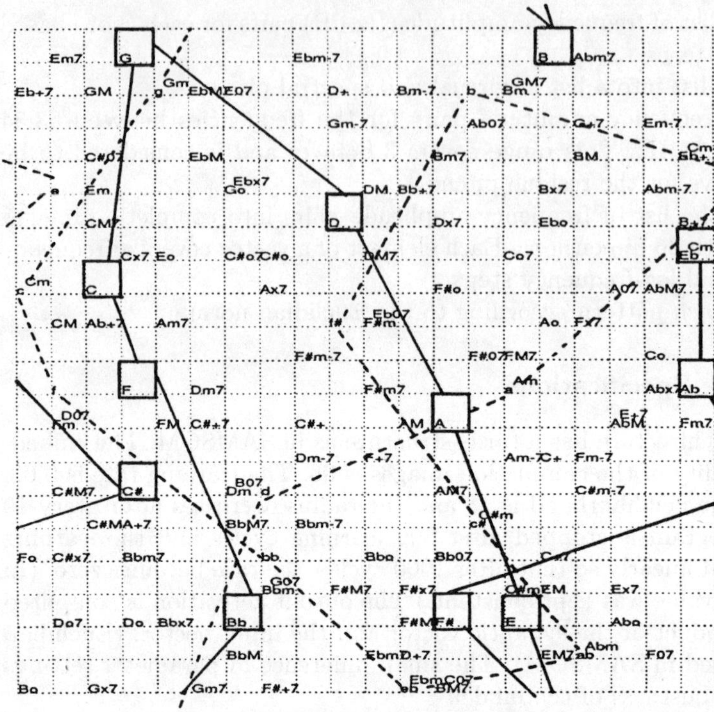

Fig. 7.8. TAMSOM map of characteristic neurons after 500 cycles

To have an idea of the relationships between these patterns (independent from the network simulation) the similarity of all tone center images with the tone center images of **C** and **c** were calculated. Afterwards, the obtained results were compared with Krumhansl's tone profiles. The results are shown in Fig. 7.9. The correlation between the two structures is 0.97 and 0.93 for major and minor tone centers, respectively.

Figure 7.8 shows the location of the tone center images and their connections. The connections are normally deduced from an analysis of the RRs but a careful observation of the location of chords might already give an indication of parts of the structure. The location of **F♯** is particularly strange because one would expect to find this CN at the other side of **E**. The connections would then be more fluent. The connections **B♭-F-C-G-D-A-E-B**, however, clearly display a part of the circle of fifths. The map of CNs for minor tone centers is even more complex and actually not very clearly distinguishable.

Figure 7.10 gives an overview of the network response structure as seen from the viewpoint of **C** and **c**. The dotted curve is compared with the curve of Krumhansl's data. A correlation of the two curves gives values of 0.91 and 0.84 for major and minor, respectively. This illustrates that the representation of the minor tone images in the network is less clear than that of the major tone images but the correlation coefficients are still highly significant.

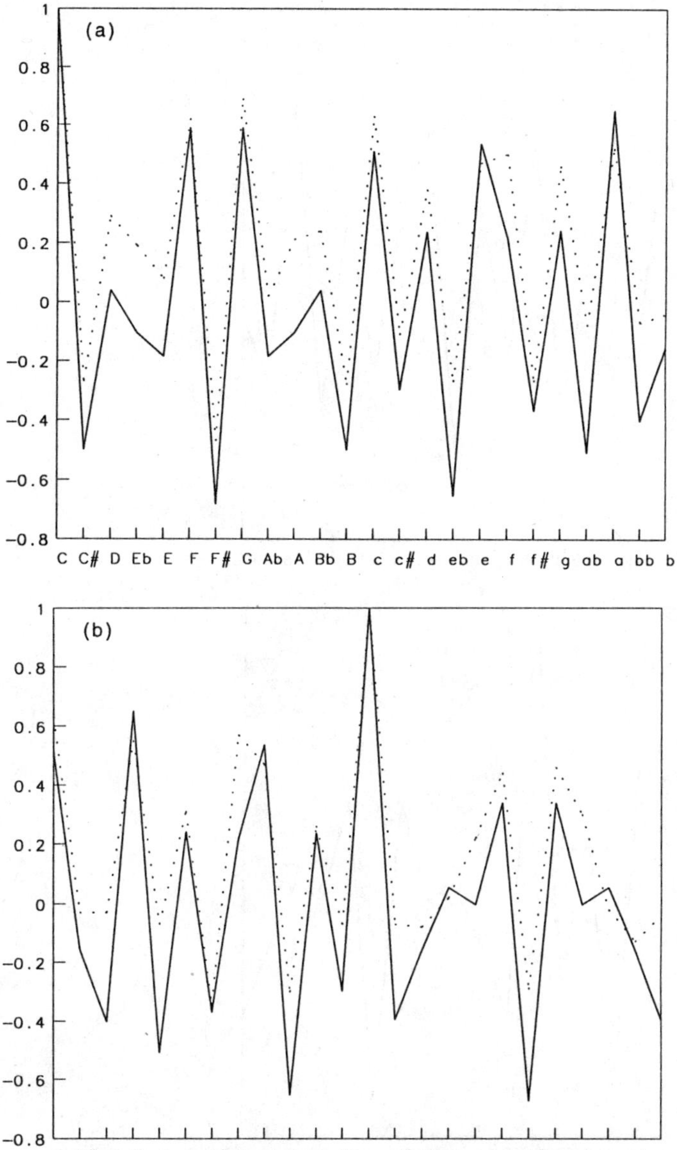

Fig. 7.9. Similarity structure of tone center images: (a) similarity structure with respect to **C**, (b) similarity structure with respect to **c**

Tone Center/Chord Relationships. The tone center/chord relationships, which provide information about stability and instability, can be illustrated by calculating the similarities between the [M]-, [m]-, and [o]-chord RRs and RR(**C**):

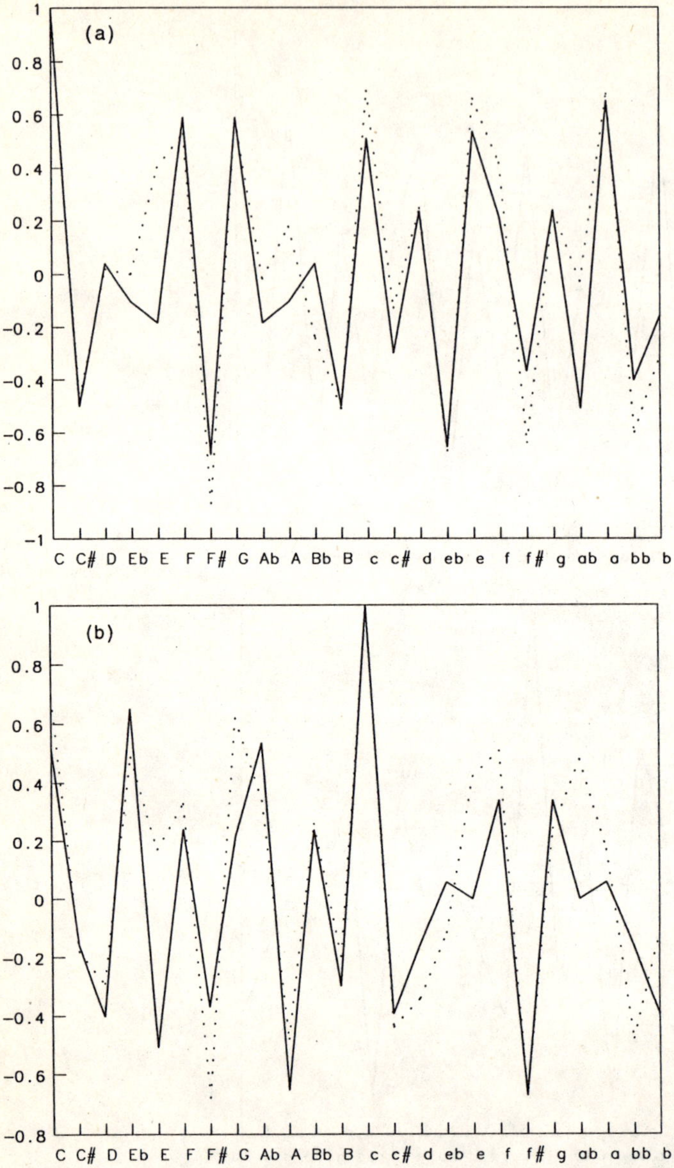

Fig. 7.10. TAMSOM network response structure to tone center images: (a) network response structure to **C**, (b) network response structure to **c**. The structures show the similarity of RR(**C**) and RR(**c**) with respect to all other RRs of tone center images

CM(0.92), C♯M(-0.35), DM(-0.28), E♭M(0.07), EM(0.68), FM(0.16), F♯M(-0.76), GM(0.47), A♭M(0.04), AM(0.17), B♭M(-0.58), BM(-0.19), Cm(0.79), C♯m(-0.08), Dm(0.15), E♭m(-0.72), Em(0.69), Fm(0.24), F♯m(-0.53), Gm(0.30), A♭m(-0.05), Am(0.56), B♭m(-0.66), Bm(-0.27), Co(-0.12), C♯o(0.41), Do(0.00), E♭o(-0.42), Eo(0.57), Fo(0.06), F♯o(-0.23), Go(0.16), A♭o(0.01), Ao(0.25), B♭o(-0.45), Bo(0.12).

7.3 VAMSOM

VAMSOM combines of the auditory model VAM and the self-organizing model SOM. The network setup is similar to the SAMSOM and TAMSOM simulations. I will not go into a detailed analysis of the self-organizing properties of VAMSOM since the results are basically similar to SAMSOM and TAMSOM. Because of the nature of VAM, the preprocessing stage is based on a leaky integration technique which is explained in the next chapter.

7.3.1 Selection of Data and Preprocessing

The training set contains 194 completion images: 115 chord images-out-of-time (described in Sect. 7.1.1), together with single tones, and intervals (two tones sounding together). The completion images are obtained by the following processing stages:

1. The Shepard-tones, intervals and chords are computed by a sound-compiler program CSOUND [7.10] (Appendix A). All 194 sound objects have a duration of 500 ms, and a short exponential attack and decay of 30 ms.
2. The sampling rate of the original signal (22050 sa/s) is converted to 20000 sa/s, in order to fit with the sampling rate of the auditory model VAM. The signal is then processed with VAM[5].
3. The tones, intervals and chords are processed as time-independent sounds isolated from each other. This is done by integrating the VAM completion images of each sound with a leaky integrator (Sect. 8.3). Out of the sequence of integrated images (called *context images*) only the last image is extracted. This image is assumed to be the representative image-out-of-time of that sound.
4. The 194 images-out-of-time are then normalized according to the Euclidian Norm.

[5] The parameters have been specified on p. 57.

7.3.2 Network Specifications

The network architecture is a torus comprising 20 by 20 neurons. The embedded dimensionality of the images-out-of-time is 56. The learning rate is 0.05, the radius of the neighborhood is 18 and it decreases every 10 cycles. The program terminates after 300 cycles. The learning rate is constant (it does not decrease) and the output is computed in terms of the Euclidian distance (like in SAMSOM). As mentioned before, these parameters are robust and can be slightly changed without affecting the global behavior of the network too much.

7.3.3 Ordering and Emergence

The error evolution of the learning process is shown in Fig. 7.11. As in SAMSOM, this error is calculated by taking the mean of the Euclidian distances between the synaptic vectors of the CNs and the corresponding images-out-of-time.

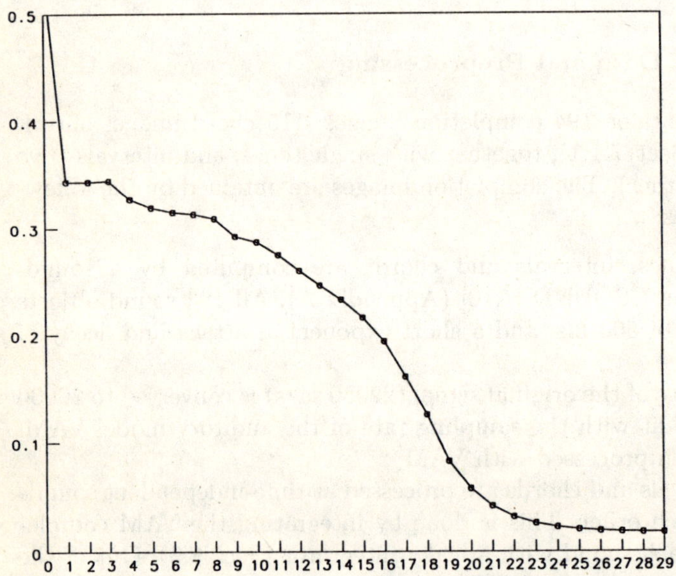

Fig. 7.11. VAMSOM error evolution over 300 cycles

7.3.4 Tone Center Relationships

The tone center images have been computed using the leaky integration technique that has been used for the calculation of the chord images. The technique is fully explained in Sect. 9.3. The basic idea is to integrate sequences of chords that form cadences and extract the last context image as a representative of the cadence. The resulting tone center images are shown in Fig. 9.3.

The network RRs to these tone center images were compared with each other, yielding the network response structure to the tone center images. This structure was then compared with Krumhansl's tone profiles structure. The correlation between the two structures is 0.84 for the major and 0.85 for the minor tone centers, which shows a significant emergent response structure to tone center images.

7.4 Conclusion

The relationships between images-out-of-time can be mapped onto a two-dimensional neural network by self-learning, that is, by adaptation to the tone image environment. The local and global properties of the functional organization in the network has a meaningful musical interpretation: chords that are similar to each other are interchangeable. Their CN is close to each other on the map and the RRs are similar. In addition, chords of the same type have response structures that suggest a global organization in circles of fifths. The network develops itself as a carrier for an accurate representation of this structure.

Tone center images provoke a response structure which is ordered, too. This is interpreted as an emergent property of the network because the patterns have not been learned. The neural carrier can develop a response structure which is similar to the response structures known from psychological research (Krumhansl's data).

A most important observation is that the distinguished images (tones, intervals, chords, tone centers) are all carried by one and the same memory structure and that the origin of such an analogical representation cannot be considered independently from the underlying dynamics. This differs with the cognitive structuralist approach, where analogical representations are often considered without reference to the dynamics.

Part of the underlying dynamics, however, depends on the preprocessing by auditory models. The tone images are imbedded in a multi-dimensional space but they have an inherent two-dimensional structure that can be mapped onto the network.

Further research is needed to figure out if the well-established pitch perception schema of musicians is a result of the fine analytical properties of the auditory system or of the self-organizing capabilities of the brain. The simulations suggest that slight deformations of the tone images may have a large influence on cognitive representations. On the other hand, the self-organizing model is quite robust: the inherent structure of chords out of time, whose image is computed by the three preprocessing devices which are being used, is preserved well on the map.

8. Learning Images-in-Time

In this chapter, tone center perception is viewed as an inherently temporal process. Long-term learning is discussed from the viewpoint of tone images-in-time.

8.1 Temporal Constraints in Tonality Perception

Butler and *Brown* [8.1] give an overview of the temporal aspects involved in tone center perception. Figure 8.1 gives one example in which listeners were asked to identify the key. Listeners were 86% accurate in identifying **D** as the tone center of the excerpt. But when the pitches of the excerpt were ordered as in Fig. 8.1b1, the agreement on **D** increased to 95%. The ordering shown in Fig. 8.1b2 promoted both **D** (41 %) and **G** (45%) as the main tone center, while an ordering as in Fig. 8.1b3 promoted **D** for only 45%. The other responses were divided over seven other pitches.

The example clearly illustrates that pitch relations alone are not sufficient to convey a sense of tone center. The perception of a tone center is largely influenced by the temporal order in which pitch relations appear. H. Brown and D. Butler explain the effect by the occurrence of rare intervals, such as the semitones and the tritone. These intervals occur infrequently within the major diatonic set so that they serve as a cue to gain a sense of tone center. Other intervals (such as the major seconds, minor thirds, major thirds, and perfect fourths) occur more frequently and are cues to multiple tone centers. The rarer the interval occurs in the diatonic set, the better it serves as a perceptual cue which the listener can use to gain a sense of tone center. Listeners thus get their tonal bearings from critical intervallic relationships arranged in conventional patterns across time.

The present computer model has no sense of intervallic cues but seems to solve the inherent temporal perception of tone center by pattern-integration over time. In the subsequent sections, the technique of pattern-integration is explained and applied in a learning task.

(a) (b)

Fig. 8.1a,b. The role of temporal order in tone center perception (Adapted from [8.1]). (a) Excerpt from Schubert, Sonata D. 664, movement III. (b) Serially ordered pitches from the excerpt

8.2 Tone Images-in-Time

Time-dependencies between images can be represented implicitly or explicitly. An ordered list of images, in which each image stands for the neuronal activation during a short time interval, is an explicit representation. The images are sampled states of the auditory system and the order reflects the musical stream of information.

Consider the chord sequence **CM-FM-GM** represented as S-patterns and assume that each chord takes only 1 s. At a sampling rate of 2 samples per second (SR = 2 sa/s), the sequence may be represented by the following list:

```
1 0 0 0 1 0 0 1 0 0 0 0 CM
1 0 0 0 1 0 0 1 0 0 0 0 CM
1 0 0 0 0 1 0 0 0 1 0 0 FM
1 0 0 0 0 1 0 0 0 1 0 0 FM
0 0 1 0 0 0 0 1 0 0 0 1 GM
0 0 1 0 0 0 0 1 0 0 0 1 GM
```

SAM transforms these patterns into the following tone completion images:

```
1.83 0.10 0.45 0.33 1.10 0.70 0.25 1.00 0.33 0.85 0.20 0.00 CM
1.83 0.10 0.45 0.33 1.10 0.70 0.25 1.00 0.33 0.85 0.20 0.00 CM
1.00 0.33 0.85 0.20 0.00 1.83 0.10 0.45 0.33 1.10 0.70 0.25 FM
1.00 0.33 0.85 0.20 0.00 1.83 0.10 0.45 0.33 1.10 0.70 0.25 FM
0.70 0.25 1.00 0.33 0.85 0.20 0.00 1.83 0.10 0.45 0.33 1.10 GM
0.70 0.25 1.00 0.33 0.85 0.20 0.00 1.83 0.10 0.45 0.33 1.10 GM
```

The images reflect the subharmonic-sum spectrum analysis of the signal in a window of 500 ms.

There is a problem, however, when such a list is used as a training set for SOM. As mentioned in Sect. 6.3, the algorithm does not consider data as ordered and works "out-of-time" – because the order of the patterns during each training cycle is randomized, so as to avoid biases. One solution would be to modify SOM by allowing temporal integration in the output of the neurons. Each neuron would then display a temporal characteristic that is defined by its impulse response. This would make the neurons susceptible to temporal information [8.6]. More recently, the Kohonen network has been modified to allow the mapping of sequences of inputs without having to resort to external time delay mechanisms [8.2].

Such a modification – however plausible from a biological point of view – has not been realized within the constraints of the present studies. The approach is restricted to a short-time integration which is external to the network. As such, the network is kept as simple as possible and temporal dependencies are represented in the images. The advantage is that integration occurs at the preprocessing level and is therefore separated from the process of self-organization.

8.3 Tone Context Images

The temporal aspects of the information flow are reflected by images which incorporate a recent history. This idea is in essence a generalization of the simple pattern integration technique used in Sect. 7.1.5. The basic requirements are threefold:

1. The duration of a tone has an effect on the construction of the context image. Short durations will contribute less than long durations.
2. The contribution of a long duration my lead to saturation. If the duration is too long, it will continue to influence the context image, but its influence will cease to grow.
3. When the tone stops being played, then its contribution to the context image must decrease. After a certain period of time the tone will be completely forgotten.

A suitable expression for the last requirement is given by the decay function

$$v_i(t+1) = \left(1 - \frac{1}{w}\right) v_i(t), \tag{8.1}$$

where w is the time window expressed in numbers of samples. This equation says that the context value of element v_i at time $t+1$ is part of the context value at time t. If $w = 1$ then the system forgets immediately. If w is large, then retention is large.

The first requirement is satisfied by allowing the context value to increase in proportion to the amplitude of the tone, such as in

$$v_i(t + 1) = v_i(t) + A_i(t), \tag{8.2}$$

where $A_i(t)$ is the amplitude of element i at time t. When the tone is not played, the amplitude is of course zero.

The combined effect of (8.1) and (8.2) also satisfies the second requirement. Indeed, when the amplitude is constant, then the context value increases in proportion to an inversed exponential. The three requirements are then satisfied by

$$v_i\,(t + 1) = \left(1 - \frac{1}{w}\right) v_i(t) + A_i(t). \tag{8.3}$$

The equation, known as a leaky integrator, can be used as a model for neuronal integration. The context value, normalized with respect to the time constant w, is called CV.

The effects are illustrated in Fig. 8.2 for a (one-dimensional) signal with a duration of 100 samples and constant amplitude ($A_i = 1$). The window is set to 20 samples and the unit is 20 CV. After 100 steps the context value reaches the unit value. When $A_i = 0$, the context value decreases gradually to zero. Applying (8.3) to a flow of completion images produces the *tone context images*.

The effect of integration on the S-pattern representation of the chord sequence **CM-FM-GM-CM** is shown in Fig. 8.3. The horizontal axis represents time, the vertical lines mark intervals of 1 s. Each chord has a duration

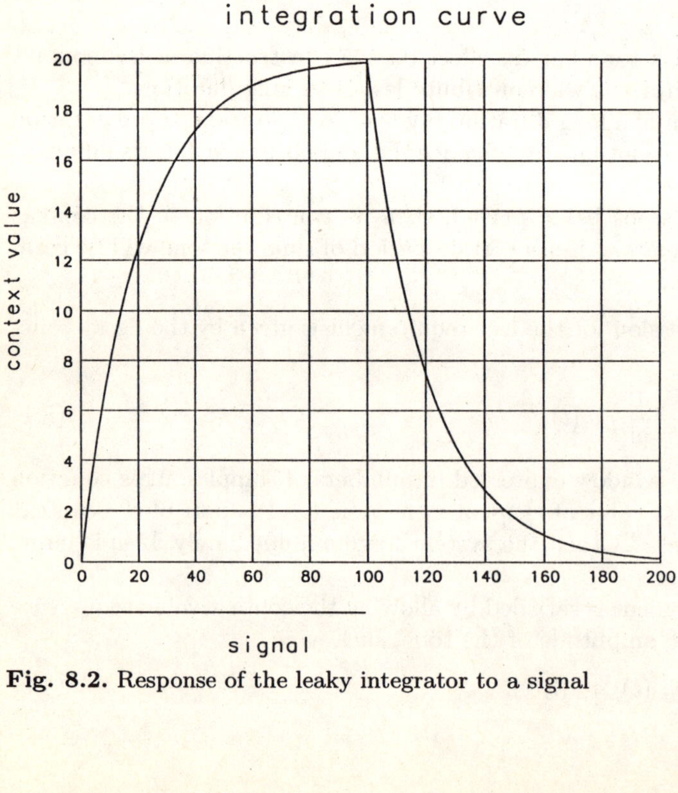

Fig. 8.2. Response of the leaky integrator to a signal

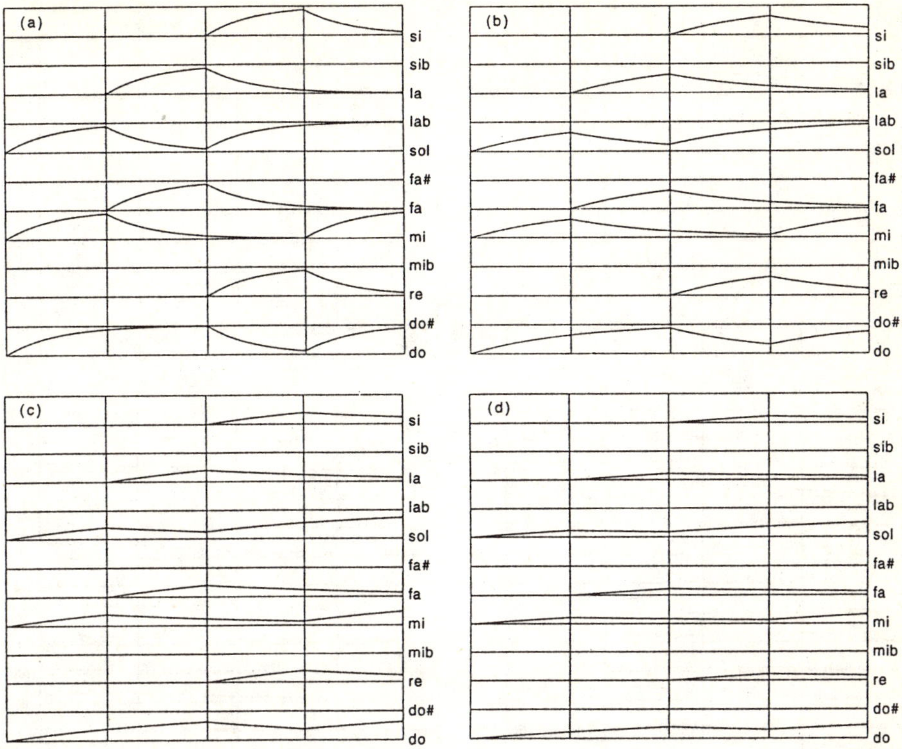

Fig. 8.3a–d. Context values of the S-pattern representation of the chord sequence **CM-FM-GM-CM**. The length of the integration window changes from 10, 20,40, to 80 samples. The horizontal axis represents the time, while the vertical axis contains the context values for each note of the chromatic scale

of 1 s and the sampling rate is 20 sa/s. In this example, the embedding space of the images is 12-dimensional and each parameter corresponds to a note label. The window is the parameter that changes over the figures: from 10 sa/s in Fig. 8.3a, to 20, 40, and 80 sa/s in Fig. 8.3. The maximum context is the unit context value (CV).

The effect of applying the integration onto the R-patterns is shown in Fig. 8.4. Euclidian normalization was applied before integration.

In Fig. 8.4a, the unit context value (CV=10) is exceeded because some inputs are greater than 1. With larger windows, the duration of the tones must be longer in order to generate a similar effect.

These figures show that an appropriate choice of the window length largely contributes to the determination of the object of interest. When the window is small (in Fig. 8.4a where it corresponds to 0.5 s) the images will mainly reflect the chords. The best image is represented at the end of the chord. When the window is larger, such as in Fig. 8.4d, the images reflect information of the temporal tone context.

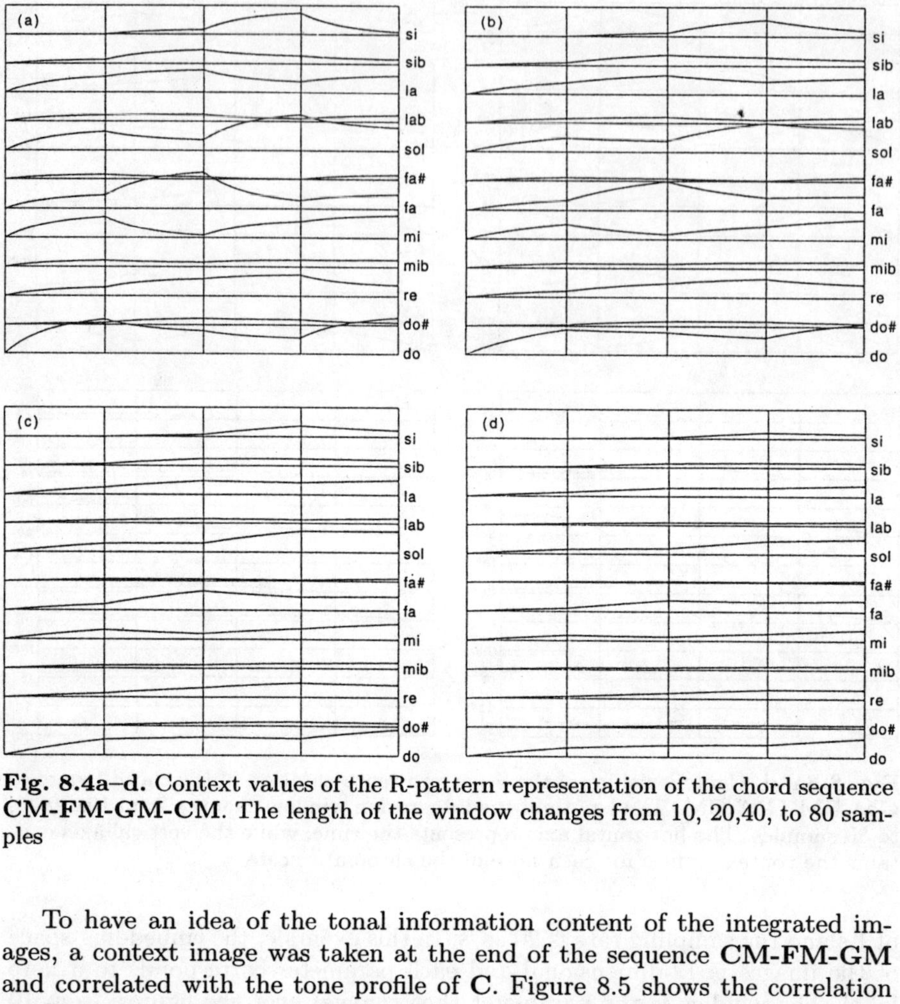

Fig. 8.4a–d. Context values of the R-pattern representation of the chord sequence **CM-FM-GM-CM**. The length of the window changes from 10, 20,40, to 80 samples

To have an idea of the tonal information content of the integrated images, a context image was taken at the end of the sequence **CM-FM-GM** and correlated with the tone profile of **C**. Figure 8.5 shows the correlation coefficient in function of the time window. The duration of the chords is the parameter which changes. The horizontal axis represents the length of the time window in samples. The sampling rate is 20 sa/s. The vertical axis gives the correlation coefficient of the context image with the tone profile.

The upper curve represents the correlation coefficients based on chord durations of 0.5 s each. The context image is therefore taken at 1.5 s. The second curve represents the correlation coefficients based on chord durations of 1 s each. The context image is then taken at 3 s. For the third and fourth curves, which have chord durations of 2 s and 4 s, the snapshot (context image) is taken at 6 s and 12 s, respectively. The correlation coefficient varies with chord duration and the size of the time window. The smaller both chord duration and time window, the higher the correlation. For long notes, a small time window scores bad results whereas a large window produces almost

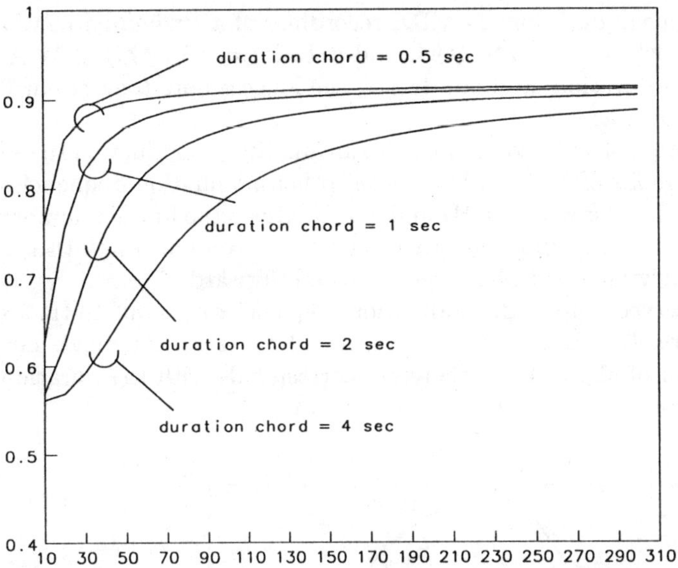

Fig. 8.5. The effect of chord duration and leaky integration on recognition. The figure shows the correlation coefficients on the vertical axis. The horizontal axis represents the length of the time window in samples. The curves show the similarity of a context image with the tone profile of **C** for different durations of the chords

equally well for long and short durations. This does not mean, however, that a very long integration time would be a good choice. Music does not consists of long or short notes only. When the window becomes too large, changes in tone center will be more difficult to follow and the notion of context will lose its meaning. Similarly, when the window becomes too small, changes in tone center will become too much influenced by long durations.

It follows from these observations that the determination of the context depends on the chosen time scale. But an appropriate time-span, one that fits the notion of tone center, will probably be one that accounts best for mean note durations.

As will be discussed in Chap. 10, integration can probably be improved by making it dependent on musical phrase. This idea, however, has not been worked out. In this book, the integration constant will be independent from the musical phrase.

8.4 Determination of the Integration Period

According to a study by *Knopoff* and *Hutchinson* [8.5], who analysed the 48 pieces of J.S. Bach's *Well Tempered Clavier* (Books I and II), 60 % of all the durations greater than 20 hs are smaller than 50 hs and about 90 % of them are smaller than 100 hs.

Measuring the mean durations in MIDI-recordings of a *Preludium (BWV 871)* and *Fuga (BWV 853)* of J.S. Bach and a *Fantasia (KV475)* of W.A. Mozart, we found a duration of about 44 hs. The longest note took about 3 s, the shortest about 2 hs.

The mean duration of 44 hs corresponds to an amplitude-modulated signal with a modulation index of 2.27 Hz. Most of the other modulation frequencies are situated somewhere between 1.5 Hz and 5.5 Hz. Given the limited number of analysed pieces, these results are provisional but a comparison with the modulation frequency of speech, however, is straightforward.

Plomp [8.7] observed amplitude modulation frequencies of about 13 Hz for phonemes, 5 Hz for syllables, 2.5 Hz for words and 0.25 Hz for sentences. The mean tone duration of about 44 hs therefore corresponds with the duration of words (100/2.5=40 hs) (Fig. 8.6).

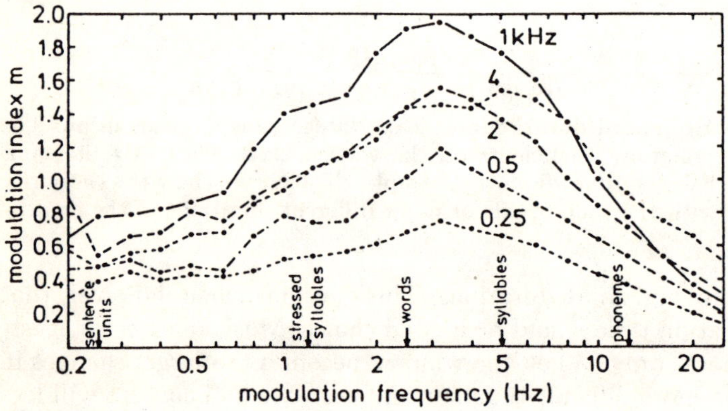

Fig. 8.6. Modulation and speech [8.7]. Speech is considered as a modulated signal. The curves represent the average spectra of a low-pass filtered speech signal (cut-off frequency 30 Hz) from one-minute discourses of ten male speakers. The envelope spectrum is largely independent of audio frequency, and has a maximum at about 4 Hz

The time interval in which context is built up is assumed to be larger. In the processing of speech it is reasonable to suppose that entities with the duration of a sentence constitute a context that may determine the context-sensitive meaning of words. The time frame of such a context window is 4 s (100/0.25=4 s). This value might be relevant for music, too – as an analysis of the psychological literature suggests. In both cases there is an interactive relationship of smaller entities (words/tones or words/chords) with the context (sentences/tone centers) in which they appear. The mutual interaction between these entities determines semantic aspects of the constituents. In language, however, the proper (or denotative) meaning of words is often more important than the context-sensitive (or connotative) meaning but context-sensitive meaning may play an important role in the resolution or creation of semantic ambiguities.

Apart from considerations about the AM frequency of meaningful entities in music and language, an appropriate choice for the time interval for image integration may be deduced from other sources – such as the relevance of different time scales in the temporal experience of humans. The following list gives an idea of different levels of integration:

– At time intervals of 3 to 5 ms, subjects can distinguish the asynchronicity of binaural acoustical stimuli [8.8].
– A time-span of 30 to 50 ms is needed to recognize temporal order. [8.9-11]. Similar values are found for asynchronous onsets of tones played by musicians. This time corresponds with the minimum time that is necessary to identify an object.
– In general, the limits at which subjective rhythmization could appear is between 120 and 1800 ms: 120 ms being the minimum interval between two rapid motor taps, 1800 ms being the time at which two stimuli are not longer linked. The optimum is 600 ms, a value that corresponds with the spontaneous (personal or mental) tempo of finger tapping, or swinging a leg or arm [8.4].
– Events are not perceived as isolated facts, but create perceptive forms (Gestalts) in which different events are combined into "pictures of the present". This so-called psychological present or precategorical acoustic storage organizes a succession of events within defined limits. Experiments show that the grouping of rhythmic patterns is typically limited to a length of about 5 s [8.4]. The subjective experience of the "present" would require an integration of subsequent data over a time-span of about 3 s on the average [8.8].

The above considerations suggest a time constant of 3 to 5 s for context images. Fig. 8.5 seems to support this constant. A window of 4 s (80 sa) yields a context image that is highly similar (a correlation value of 0.85) with the tone profile of **C**, even if the chord durations extend to 2 s each. At long durations (4 s for each chord) the image of a global tone center for the sequence is lost.

8.5 TAMSOM

This section evaluates the learning of images-in-time with TAMSOM. In the ideal case, the data for this simulation would consist of several musical pieces from different periods and different genres but the amount of data and computer power involved would be large. For technical reasons this study is restricted to a more modest approach.

8.5.1 Selection of Data and Preprocessing

The training data are based on an example in Fétis's textbook of harmony theory [Ref. 8.3, p. 45–46]. The example is shown in Fig. 8.5.1 for the keys of **C** and **a**.

A MIDI-encoding of both pieces has been translated into S-patterns at a sampling rate of 30 hs per sample. The two files, each containing 72 S-patterns are then transposed to all other keys by vector rotation. This gives a total amount of 1728 S-patterns ($24 * 72$).

The training data are processed by TAM and the context images are obtained by a leaky integration of the completion patterns using an integration window of 3 s. With a sampling interval of 30 hs, the integration time corresponds to 10 such patterns (The unit is 10 CV)[1]. The patterns are labeled by marking all time-dependent tone images of a certain tonality with the label of

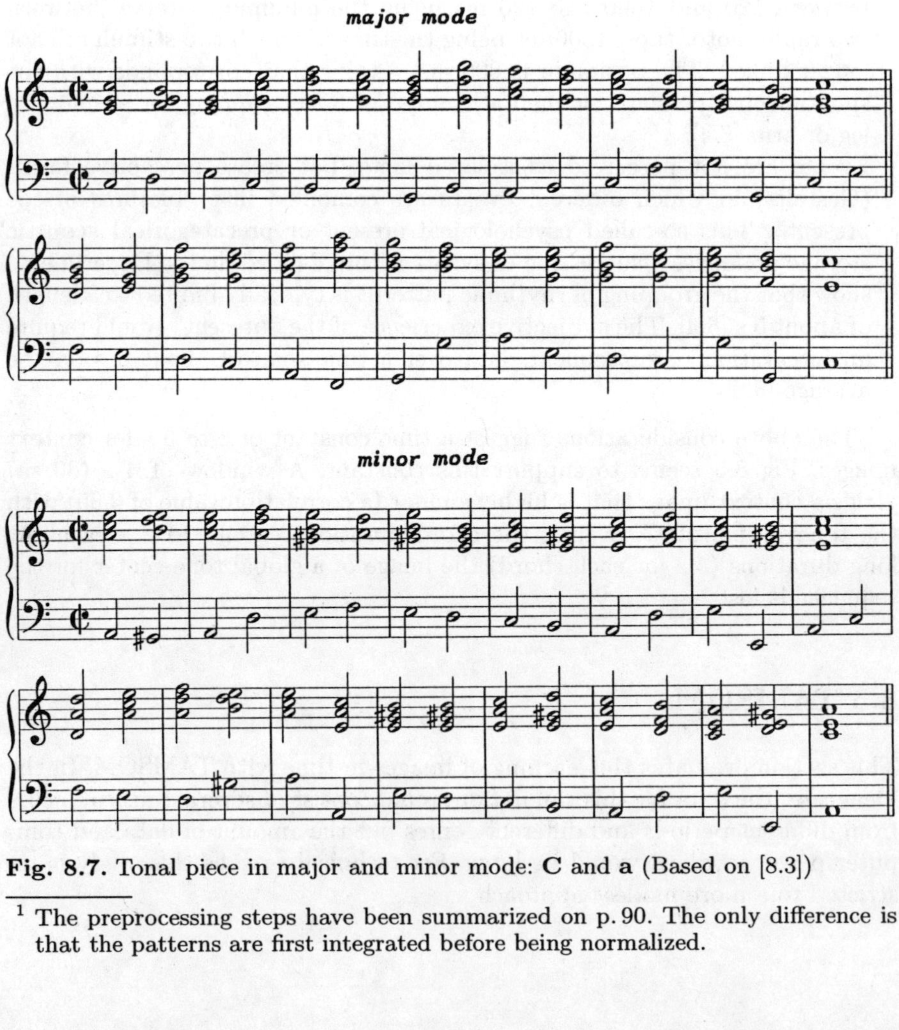

Fig. 8.7. Tonal piece in major and minor mode: **C** and **a** (Based on [8.3])

[1] The preprocessing steps have been summarized on p. 90. The only difference is that the patterns are first integrated before being normalized.

the tone center. In total, there are 24 such labels corresponding to 12 major and 12 minor tone centers.

8.5.2 Network Specifications

The network architecture is the same as the one used in Sect. 7.2.2. The learning rate is $0.02 * (1 - i/100)$, where i is the number of cycles. The radius of the neighborhood starts at 18 neurons and decreases with every cycle. Some redundancy in the training data is assumed so that the total amount of training cycles can be reduced to 50. The output activation is computed on the basis of the Euclidian distance (the same measure that is used for the adaptation of the synaptic values) and the values are normalized in that the highest and lowest are rescaled within the range of 0 to 1.

8.5.3 Aspects of Learning

Figure 8.8 shows the evolution of the mean error. More cycles would be needed to decrease the error further, but owing to the total amount of data it is of course impossible to obtain a mean error of zero, since multiple patterns cluster into a single neuron.

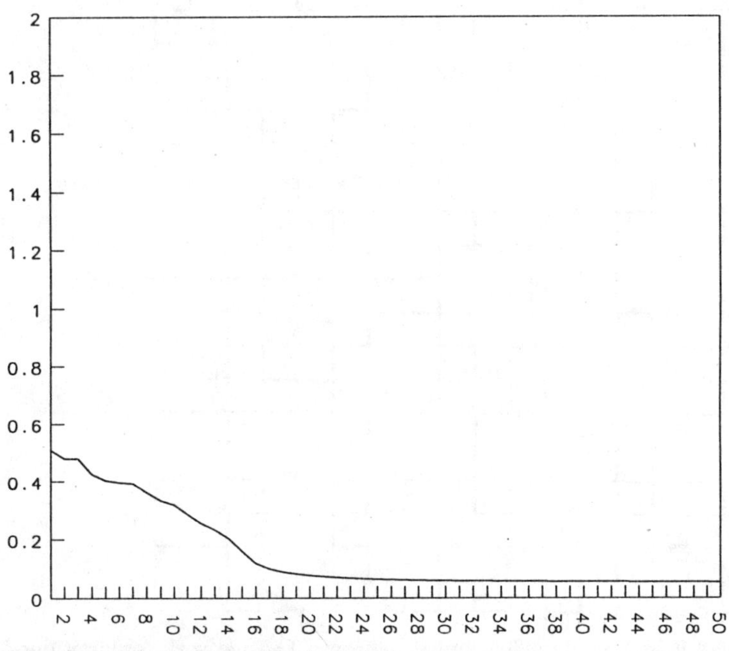

Fig. 8.8. TAMSOM evolution of the mean error

8.5.4 Aspects of Ordering and Emergence

In the previous chapter, the evaluation was based on tone center patterns (obtained by the integration of chords in cadences) or tone profiles (obtained by psychological experiments). The latter method cannot be used when the dimensionality of the auditory model is greater than 12 and the first method is restricted to the study of tone center patterns.

A third possibility relies on an evaluation of the data structures in the neural network by classifying neurons of the same class into groups. The labeled training patterns are presented to the network and the characteristic neurons (CNs) are monitored. By majority voting, the neuron that does occur most often as CN for a particular class is selected. This neuron is called the *voted characteristic neuron* (VCN).

Once this is done, the complete set of neurons is divided into different groups according to a nearest-neighbor comparison. The nearest-neighbor comparison is based on the idea that neurons which are closest in distance to a VCN get the label of that neuron. This is achieved by comparing the

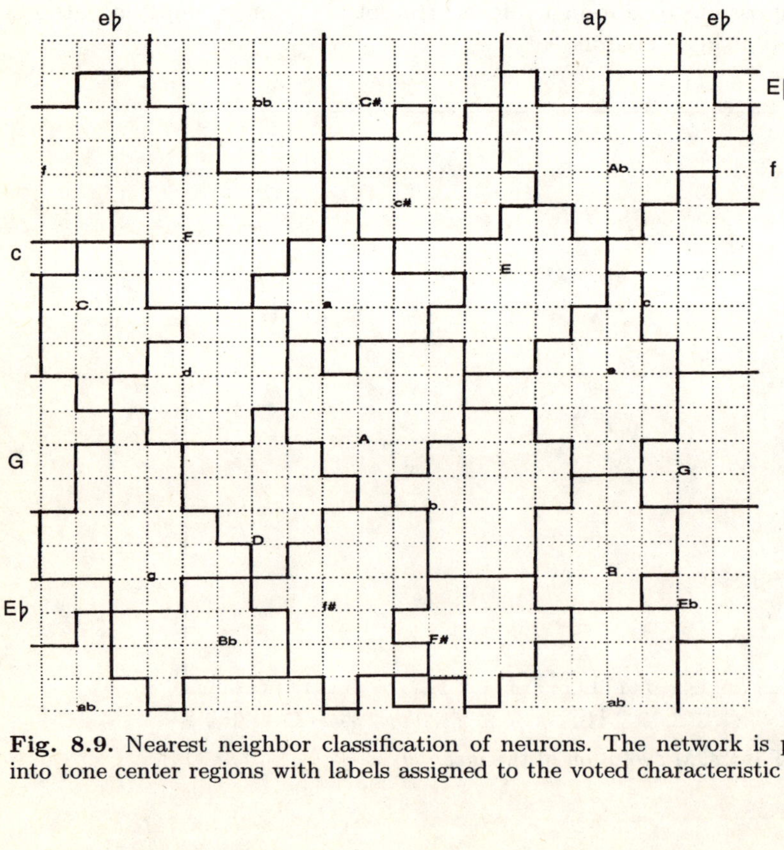

Fig. 8.9. Nearest neighbor classification of neurons. The network is partitioned into tone center regions with labels assigned to the voted characteristic neurons

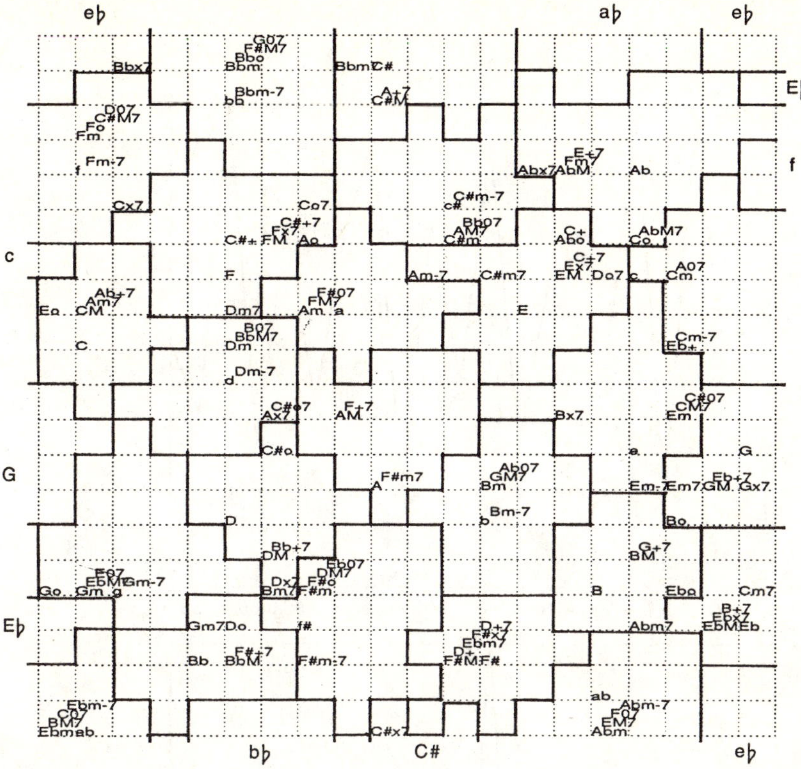

Fig. 8.10. Map of characteristic neurons to TAMSOM chords out-of-time

synaptic vector of each neuron with the synaptic vector of each VCN and assigning to that neuron the label of the most similar VCN.

The result is shown in Fig. 8.9 where the network is partitioned into tone center regions with labels assigned to the VCNs. The labels define the class membership of the neurons in that region. The network appears to be highly ordered and the circle of fifths can clearly be distinguished.

The network response to TAMSOM chord images-out-of-time and TAMSOM tone center images-out-of-time (used in Sect. 7.2) is shown in Fig. 8.10 The tone center images-out-of-time are correctly classified (recognized), as well as the most stable chord images-out-of-time – such as the major triads and the minor triads. Other chord images are more difficult to classify because they are less stable without the specification of a context. The dominant seventh chords and major seventh chords provide good examples of this ambiguity. **Cx7** is classified as a chord of **f**, while **Bx7** is classified as a chord of **e** and **Ax7** as a chord of **d**. **Gx7**, however, is classified as a chord of **G**.

The analysis of the response structure, based on the tone center images-out-of-time, is given in Fig. 8.11. The comparison with the Krumhansl curves gives correlations of 0.93 for major centers and 0.81 for minor centers.

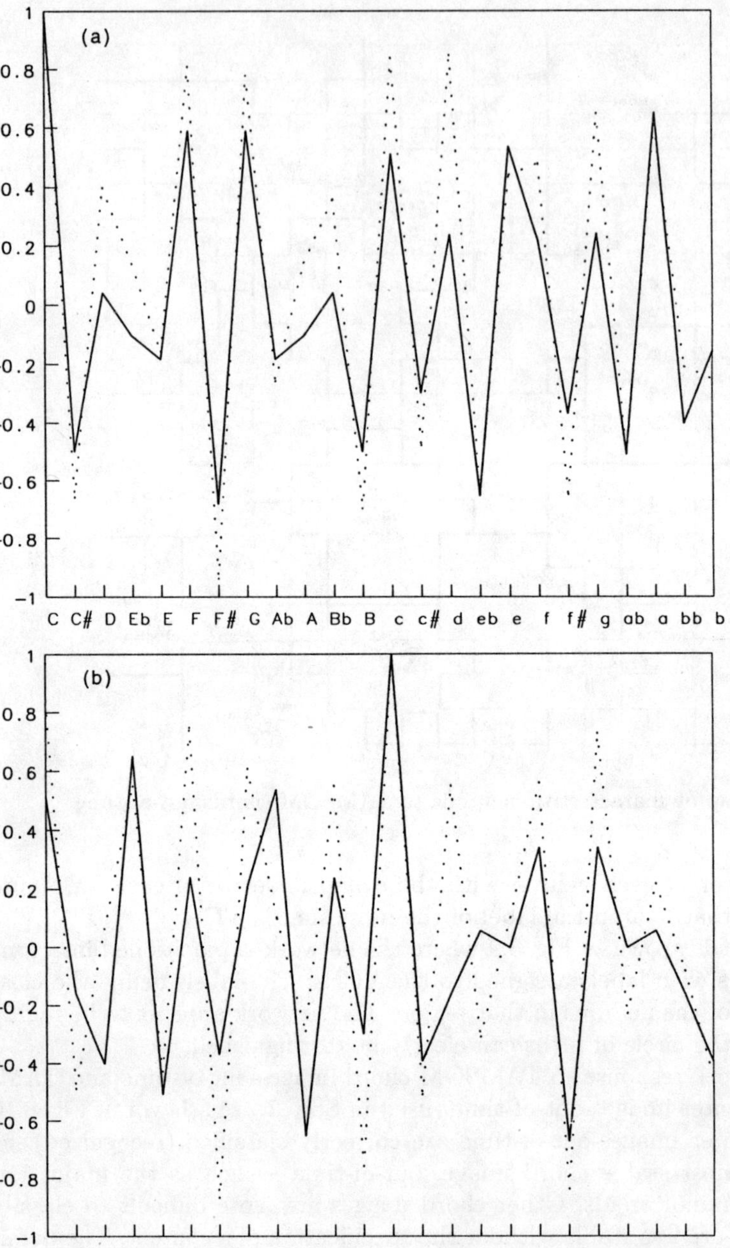

Fig. 8.11. TAMSOM network response structures of tone center images: (a) network response structure with respect to **C**, (b) network response structure with respect to **c**. The structures show the similarity of RR(**C**) and RR(**c**) with respect to all other RRs of tone center images

8.6 VAMSOM

The last simulation in the series concerned with learning is based on a large set of data. The simulation shows the power of the neural network to structure and generalize. The results are better than all previous simulations.

8.6.1 Selection of Data and Preprocessing

The training set consists of 18792 context images which represent 72 cadences. The total amount of time of the sound examples is about 3 minutes. The procedure to obtain these data involves the following steps:

1. The calculation is based on three cadence types with chord degrees **I-IV-V7-I**, **I-II-V7-I** and **I-VI-V7-I**. The chord sequences are expressed in the major and minor scale and transposed over the chromatic scale. This gives a total of (3 * 2 * 12 =) 72 different cadences.
2. The sounds have been based on Shepard-tones (Appendix A). Each chord has a duration of 500 ms and has a short exponential onset and offset of 30 ms. Between the chords, there is a rest of 200 ms.
3. All 72 cadences have been processed with VAM in a way similar to the example given in Sect. 5.5.3. VAM outputs 100 images per second and produces 261 completion images for each cadence. This gives a total of $(72 * 261 =)$ 18792 images. Each image is represented by 56 data points.
4. The context images for each cadence are obtained by a leaky integration of the VAM completion images using (8.3) ($w = 3s$). These patterns serve as the input to SOM. Figure 9.2 shows the tone context images of one cadence.

8.6.2 Network Specifications and Aspects of Learning

The architecture of the network is similar to the previous ones, and consists of a grid of 20 by 20 neurons in a torus structure. The learning rate is $0.02 * (1 - i/100000)$, where i is the number of cycles. The radius of the neighborhood starts at 18 and decreases every learning cycle. (It is assumed that the training set contains much redundant information.) During one learning cycle all 18792 data are presented to the network. The training of the network is stopped after 30 cycles. The output activation is computed on the basis of the Euclidian distance and the output values are normalized.

8.6.3 Aspects of Ordering and Emergence

The evolution of the mean error (not shown here) is similar to the previous studies. The analysis of the response structure is based on the 24 images which are extracted from the cadences. The last context image of Fig. 9.2 is considered to express the tone center of the cadence and can be use as a test

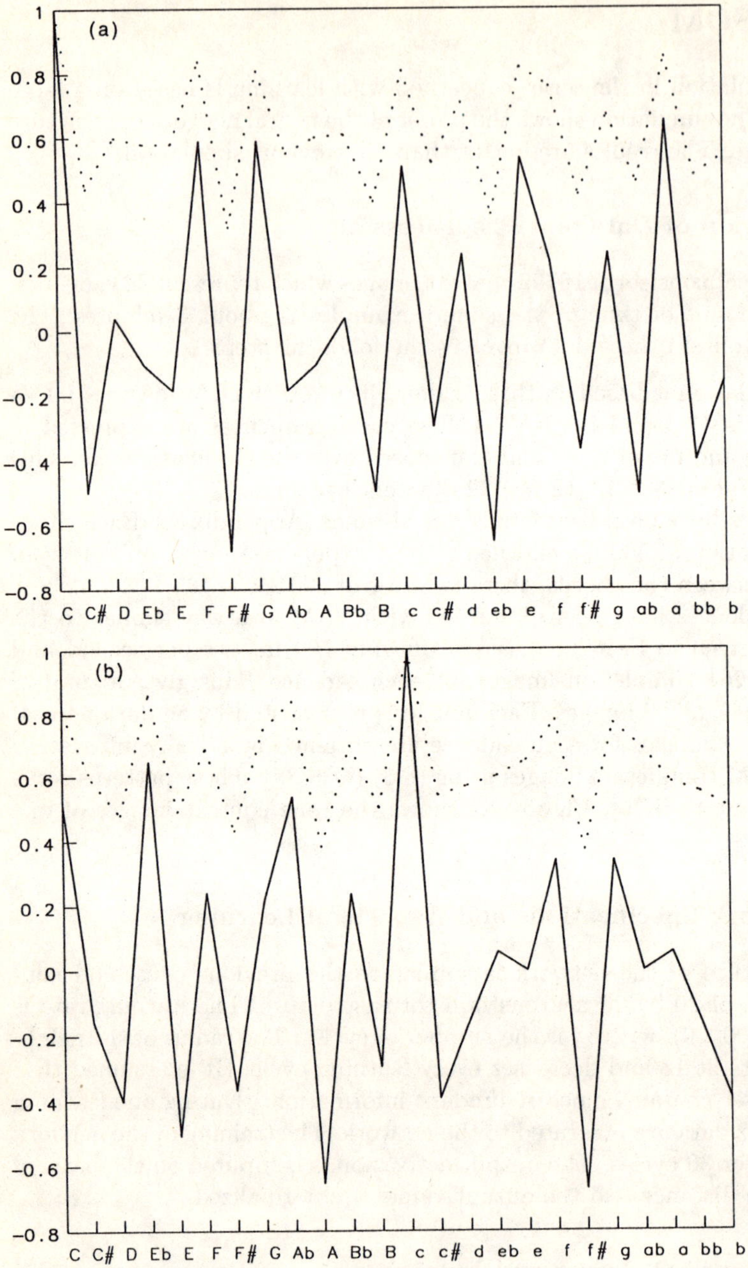

Fig. 8.12. VAMSOM network response structures of tone center images: (a) network response structure with respect to **C**, (b) network response structure with respect to **c**. The structures show the similarity of RR(**C**) and RR(**c**) with respect to all other RRs of tone center images

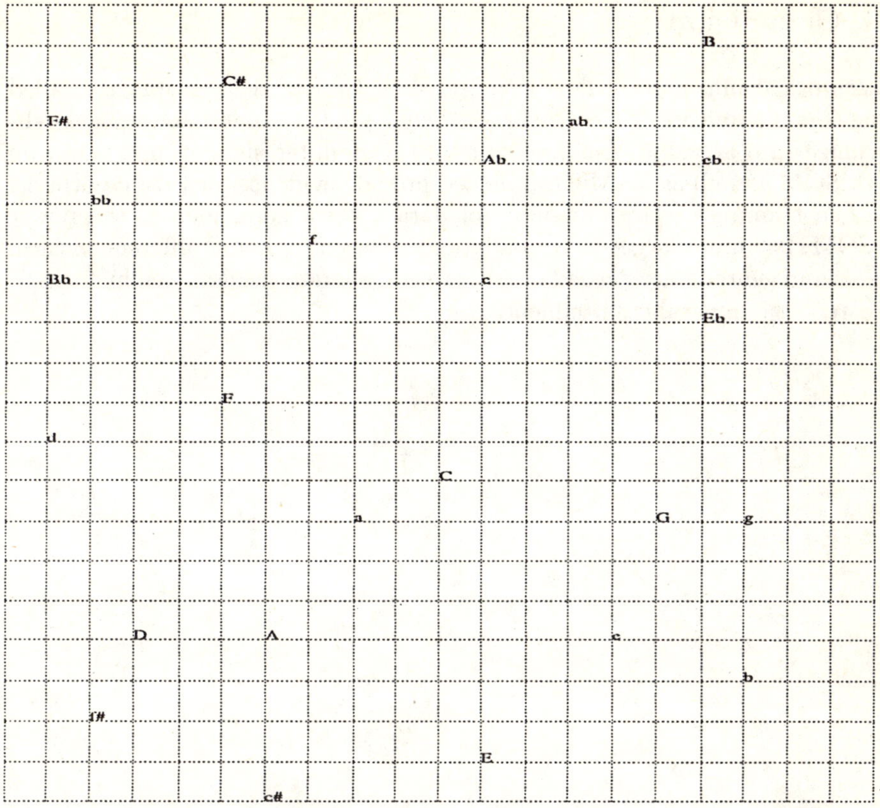

Fig. 8.13. VAMSOM map of characteristic neurons of tone center images. The circle of fifths is clearly represented in this map

pattern. This image is selected and the set of 72 such patterns is then further reduced by taking the mean of the images that represent the cadence types. This yields a list of (72 / 3 =) 24 tone center images (Fig. 9.3).

For each image, the response region (RR) (or output activation of the network neurons) is stored as a vector of 400 elements. Afterwards these vectors are compared to each other using the similarity measure of the correlation coefficient. The results are shown in Fig. 8.12. The correlation of the dotted curve (which represents the results of the model) with the full-line curve (which represents the psychological data of Krumhansl) is almost perfect: 0.99 for the major tone centers, and 0.98 for the minor tone centers. These results are better than any other results thus far obtained.

The almost perfect match with the psychological data is reflected in a nice ordering of the characteristic neurons on the map. This is shown in Fig. 8.13 where the circle of fifths can be very clearly distinguished on the torus. (We leave it to the reader to draw the lines of connection between the tone centers).

8.7 Conclusion

The musical information flow is reflected in time-integrated images, called tone context images. Tone semantics theory predicts that these images self-organize into a stable response structure. Computer simulations, based on TAMSOM as well as VAMSOM, indeed provide evidence for this hypothesis.

The training of large amounts of data gives a good idea of the power of SOM, but more large-scale simulations – based on many different musical pieces and genres – are needed in order to corroborate the assumed hypothesis in a realistic musical environment.

9. Schema and Control

This chapter is about schema-based control of perception. It introduces a model of tone center attraction dynamics which implements self-organization as an associative process.

9.1 Schema-Based Dynamics

A schema adapts to the environment and controls perception. The theory of tone semantics is based on a distinction between a *long-term data-driven* dynamics and *short-term schema-driven* dynamics.

1. **Long-Term Data-Driven Dynamics**. The long-term dynamics is responsible for the organization of neuronal functions into a stable tone perception schema. The computer simulations in Chaps. 7–8 suggest that such a functional organization may emerge just by listening to music. The long-term dynamics is *data-driven* because no domain specific knowledge or information from higher memory levels is needed.
2. **Short-Term Schema-Driven Dynamics**. The short-term dynamics is responsible for recognition and interpretation of continuously changing auditory images. To make any sense, the images reflecting the information flow (rapidly changing patterns that come from the senses) must be related to more stable memory information (which are less vulnerable to rapid changes). Hence, the dynamics is *schema-driven.*

Although both processes are supposed to operate in parallel, they are here separated to get a clearer view on the features involved. The mutual interaction – for example, how short time dynamics may influence long-term learning – is beyond the scope of the current model and is not dealt with here.

9.2 TCAD: Tone Center Attraction Dynamics

Schema-based control of perception is modelled by an attractor dynamics which implements self-organization as an associative process. The model describes association in terms of attractors, stable states and transitions towards

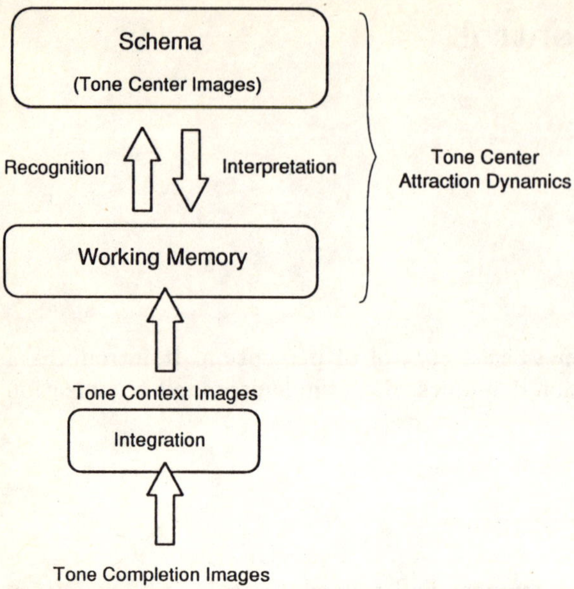

Fig. 9.1. Architecture of TCAD

stable states [9.1, 2]. It is not implemented as a neural network, but it simulates the behavior of a self-organizing associative dynamics by means of a computational algorithm – like SOM.

TCAD contains an internal dynamics which is driven by tone context images (Fig. 9.1). The working memory is a short-term buffer of a few seconds, acting like a shift-register. In the buffer, tone context images are stored and adapted by the schema.

The schema can operate in two possible ways:

- passively – when the schema is merely used as a template to match incoming images,
- actively – when the schema is actively involved in the matching process.

The first is called *TCAD-recognition*, the second is *TCAD-interpretation.* TCAD introduces a metaphor of attraction dynamics. The relation to schema theory can be clarified by considering how TCAD interprets schema responses and images.

9.2.1 Schema Responses as Semantic Images

In the previous chapters, a schema was interpreted in terms of SOM, a neural network carrier in which the response structure has analogical and topological properties. The schema response to input was called a response region (RR), and the highest activated neuron in the map was called a characteristic neuron (CN). A special case is the *voted CN*, which is the neuron which occurs most often as the CN of a particular class of inputs. The schema response,

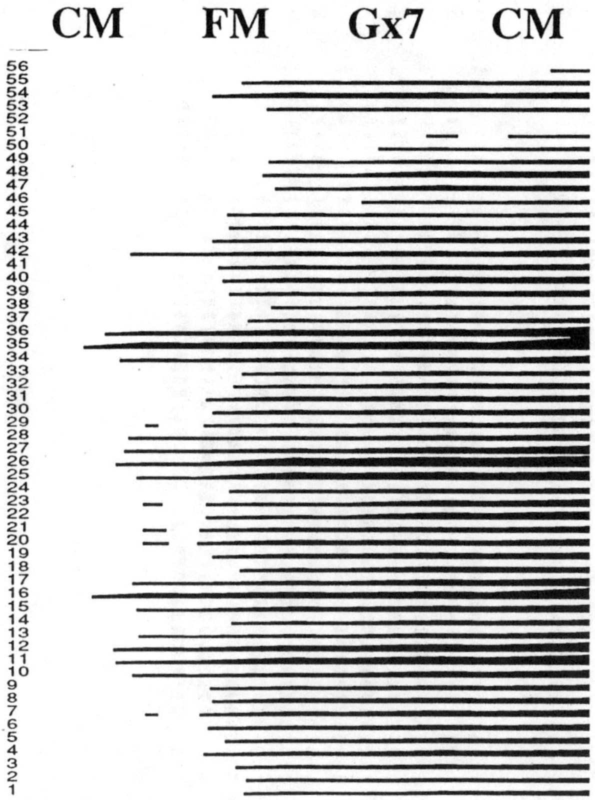

Fig. 9.2. Tone context images of the Shepard-chord sequence **CM-FM-Gx7-CM** (based on Fig. 5.11)

monitored at particular CNs, led to the notion of a characteristic response region (CRR).

TCAD assumes that the model's sense of tone center can be measured by monitoring the schema responses at the CNs of tone center images. TCAD thus reduces the schema response to a CRR, the CNs of which correspond to tone center images. In TCAD, this particular CRR is represented by a 24 dimensional vector and the values correspond to the activation of the CNs for tone centers. This particular schema response is called a *semantic image.* Semantic images convey sufficient information to indicate the model's sense of tone center in a particular piece of music.

9.2.2 Images as States

In TCAD, images are interpreted as states or points in an N-dimensional embedding space. For example, in a simple network with only 3 neurons, and binary activation (0 or 1), there are $2^3 = 8$ possible states. Each state can be represented by the vertex of a cube. The state-space is 3-dimensional. A carrier of 12 neurons with binary coding gives $2^{12} = 4096$ possible states,

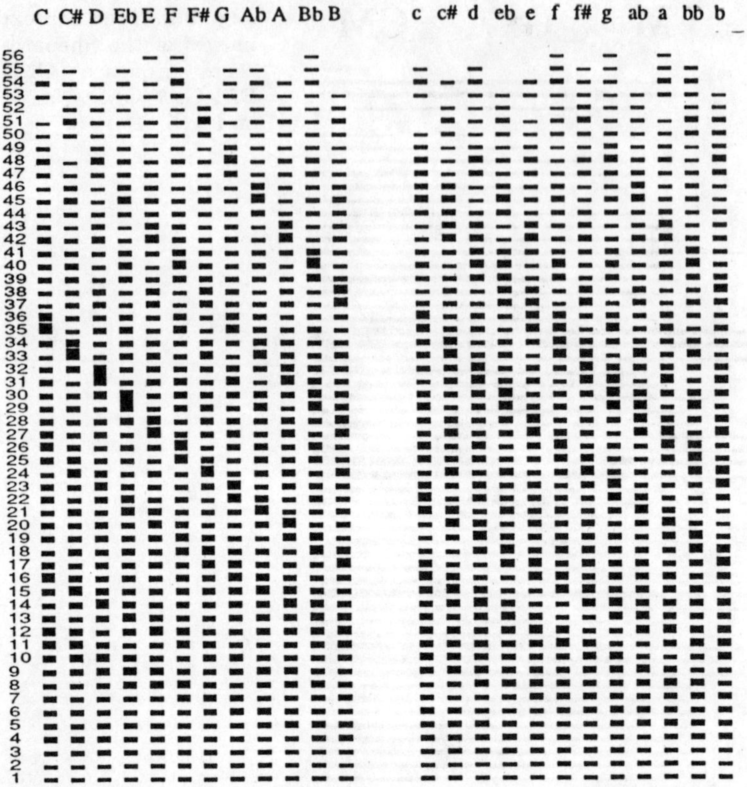

Fig. 9.3. The TCAD-stable states: tone center images obtained by pattern integration over time

corresponding to the vertex of a hypersphere. With continuous activation values, instead of binary values, the number of possible states is in principle unlimited.

States are updated at distinct time intervals and the vector values correspond to the probability of neuronal firing during that time interval. Tone center images are interpreted as stable states that attract other less stable states – in particular the tone context images. The basic principle is very simple: unstable states of the system are attracted by stable states so that after a while, the system relaxes into the stable state which was closest to the unstable state. The relaxation process, however, is continuously influenced by new unstable states (called fluctuations) which may cause transitions from one stable state to another. The properties of attraction (whether attraction is strong or weak) depend on the distances between the states. When an unstable state is close to a stable state, it will be attracted more than when it is at a greater distance.

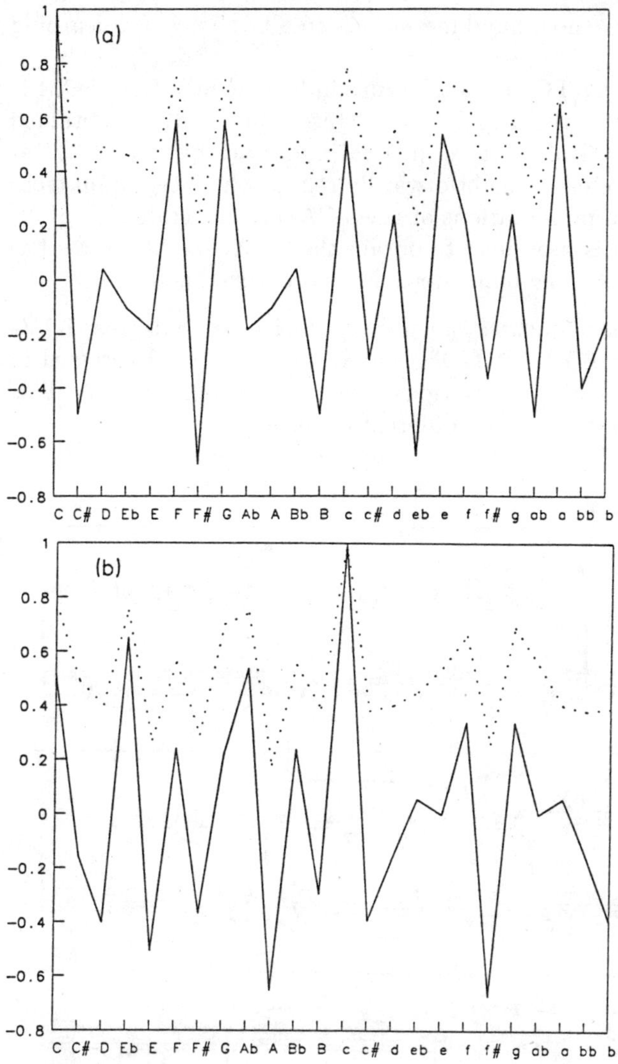

Fig. 9.4. Similarity structure of VAM-based tone center images: (a) the similarity structure is compared with tone profiles for **C**, (b) the similarity structure is compared with tone profiles for **c**

9.3 TCAD – Stable States

TCAD could in principle be built on top of SOM. For example, one could train a SOM and use the *voted characteristic neurons* (VCNs) as representations of the TCAD-stable states. The CRR would then be made up by VCN responses.

The present approach, however, discards the complexities of a neural network and reduces the schema to a table of tone center images. One could use the VCN vectors and put them in such a table, but there are alternatives. As this section presents TCAD independently from SOM, the stable points have been calculated by a short-hand method as an alternative to a learning process.

One of the reasons why TCAD can operate independently from SOM is that the representations of the TCAD-stable states can easily be obtained without neural networks. Stable states in tonal music are generated by cadences and one may therefore straightforwardly rely on the integration technique to obtain reliable representations of the TCAD-stable states.

The procedure which is used here to obtain the TCAD-stable states has been described in Sect. 8.6. The main steps are summarized here:

1. The calculation is based on three cadence types with chord degrees **I-IV-V7-I**, **I-II-V7-I** and **I-VI-V7-I**. The chord sequences are expressed in the major and minor scale and transposed over the chromatic scale. This gives a total of (3 * 2 * 12 =) 72 different cadences.

Fig. 9.5. Score excerpt of *Arabesque No. 1* (C. Debussy)

2. Since TCAD-stable states should act as prototype patterns, the sounds have been based on Shepard-tones.
3. All 72 cadences have been processed with VAM.
4. The context images for each cadence are obtained by a leaky integration of the VAM completion images (Fig. 9.2).
5. The last context image of Fig. 9.2 is considered to express the tone center of the cadence.
6. The set of 72 context images is further reduced by taking the mean of the images that represent the cadence types. This yields a list of (72 / 3 =) 24 tone center images, called TCAD-stable states.

The TCAD-stable states are shown next to each other in Figure 9.3. The vertical axis specifies the frequency in terms of the time-lags. The similarity structure of these VAM-based tone center images (Fig. 9.4) shows a correlation coefficient of 0.98 and 0.96 with the structure of the (Krumhansl) tone profiles. This is an almost perfect match which can be related to the response structure of the network (Sect. 8.6). Recall that the training of the network involved 18792 56-dimensional patterns and it took a long time to calculate the results. The synaptic vectors of the characteristic neurons of the tone center images in the trained network differ only slightly from the images displayed in Fig. 9.3. We may therefore conclude that the shorthand method is effective – both in calculation and quality of the representation.

9.4 TCAD – Recognition

The TCAD-recognition module determines the position of a context image in the framework of TCAD-stable states by measuring the distance in terms of correlation coefficients. From that viewpoint, TCAD-recognition is similar to the classical template-matching approach, in which the matching results are stored as distributed representations.

TCAD-recognition was applied to *Arabesque No. 1* by C. Debussy in a recording of T. Vásáry[1]. The score excerpt is shown in Fig. 9.5.

The audio signal was sampled at 44.1 ksa/s. It was then down-sampled to 20 ksa/s in order to fit with the sampling rate of VAM. The auditory nerve images, completion images and context images were calculated as in Sect. 5.5.3. Figure 9.6 shows the completion images at the beginning of the example.

In TCAD-recognition, the context images are compared with the 24 TCAD-stable states. Figure 9.7 displays the result of the pattern-matching. The horizontal axis is the time. The vertical axis shows the correlation coefficients for each tone center. The black strips show the tone center which has the highest value. The vertical lines mark sections of 7.5 s.

[1] CD: Deutsche Grammophon 429 517-2, 1970.

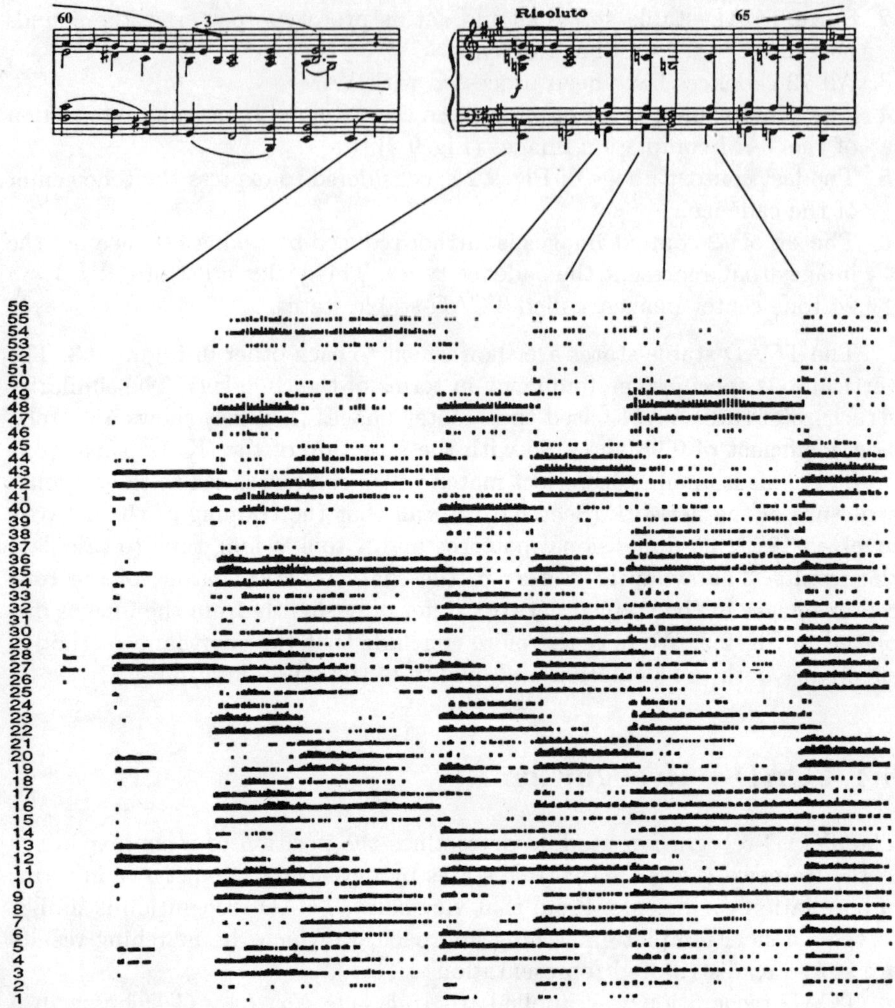

Fig. 9.6. VAM-based completion images of *Arabesque No. 1*

An important point to keep in mind is that the TCAD-stable states are based on Shepard-tones. As opposed to this, the musical signal contains piano tones. Further, it may be interesting to note that the analysis is not very different from an analysis of the same piece but based on TAM instead of VAM [9.3].

Globally speaking, the analysis suggests that the first half of the piece is in the tone center of **C**, while the second half is somewhere between **c♯**, **E** and **A**, and then in **f♯** and in **E**. The short black strips in the middle reflect the modulation.

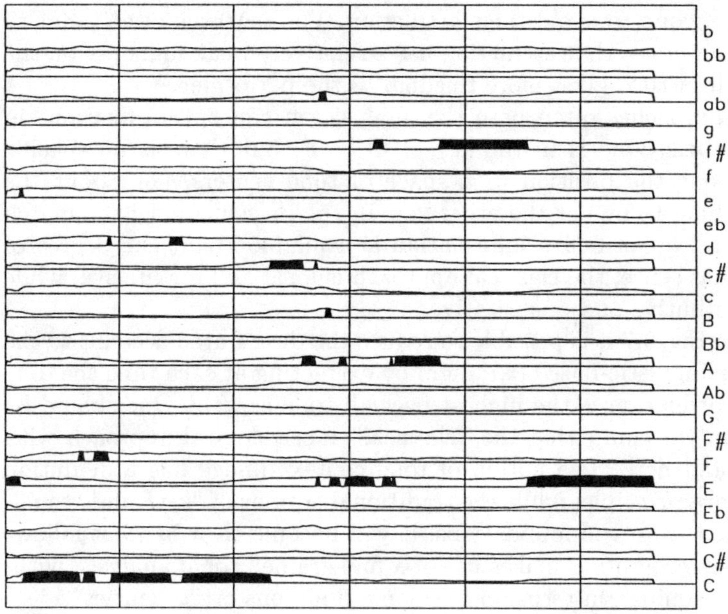

Fig. 9.7. Semantic images based on TCAD-recognition showing the similarity of the tone context image with each tone center. The black strips show the tone center which is most similar to the tone context image

A more detailed analysis of the first section (from 0 to 7.5 s) shows that in measure 65 the evidence for **F** is higher than for **C**. Most musicologists would probably argue that the first section is in **C**, without any modulation to **F**. The model is quite sensitive to the occurrence of the chords **AM**, **FM**, **CM**, **Dm7**. In the second section (7.5 to 15 s) the black strip suggests the tone center **d** for a very short period, but the main part of this section is in **C**. This continues in the third section (15 to 22.5 s) although there is a lot of movement in the tonal space. At the end of the first bar (which corresponds with in the middle of this section) there is a lot of evidence for **c**♯. At the beginning of the Primo Tempo, there is no stable high value. In the fourth section there is evidence for **E**, **A** and **f**♯. In the fifth section (30 to 37.5 s), there is a clear shift from **f**♯ to **E** at the beginning of measure 76 (a Tempo).

The sensitivity to changes in tone centers is due to the integration time-constant, which is currently set at 3 s. If this constant would be larger, then the judgments would be less susceptible to chord changes. On the other hand, a short integration time, and thus a high sensitivity of the chords, reflects a common practice in Jazz performance.

The distributed representation in Fig. 9.7 fits very well with the fact that the perception of tone center is often ambiguous. The perceived key can be related to different tone centers at the same time. In Jazz music, for example, the harmonic and melodic patterns often have no pronounced tone center and it is part of the game to avoid their attraction. The patterns often point to multiple tone centers so that attraction has a relatively weak influence on the percept – which in turn gives more freedom to the performer.

Symbolic approaches often base the analysis on conceptual fixations in terms of key labels. One is in the key of **C** or **F**, but there is no specific information about the position in a space of tone centers, nor about the degree of matching to one or the other key. In the present model the match is monitored and quantitative information is available about the degree of matching. In that sense, the classical approach is local and qualitative, while this one is distributive and quantitative.

As shown in Fig. 9.7 it is possible to reduce the distributed account to the fixations of the linguistic-based paradigm by extracting at each time the tone center whose correlation is the highest over all tone centers. One should be careful, however, in comparing the "classical" musicological approach with the one presented here. The notion of tone context image has a definition based on auditory principles while the traditional notions of "key" and "tonality" have music theoretical foundations. It will become clear in what follows that tone center recognition makes use of a fine-grained tonal analysis, without explicit "reasoning" in terms of harmonic functions or "tonalities".

9.5 TCAD – Interpretation

The idea that a schema may actively control the perception of a particular incoming image is related to the context-sensitive approach. The role of an active schema is particularly relevant in cases where previous semantic images are reconsidered in the light of new evidence.

Consider a sequence containing the chords IV-V-I. After hearing the first chord, the tone center will point to the tonic that corresponds with degree IV. It is only after hearing the rest of the sequence that the first chord can be interpreted in terms of its subdominant function. The schema should thus control the matching process and adapt the semantic images in view of new evidence.

The TCAD-recognition part is limited to a passive registration of the position of the context images in a framework of tone centers images. It does not involve any active participation of the schema. Therefore, problems such as those related to the interpretation of the IV-V-I sequence, are not resolved. Also, at the beginning of Fig. 9.7, the **MI** is typically not recognized as a part of the tone center of **C**.

TCAD should therefore control the semantic images of a recent past in view of new evidence. A first approach was conceived as follows: When a context state comes near to a particular TCAD-stable state, then it is attracted by that state. The force of attraction depends on the distance between both states. A short distance implies a great attraction force. A long distance implies a less great or non-existent attraction force. Attraction will have the

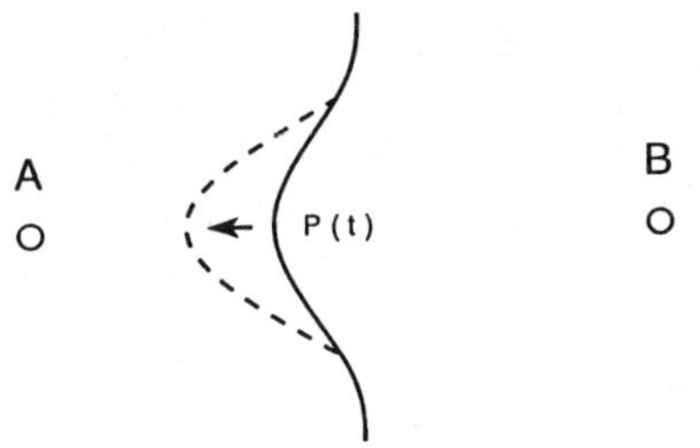

Fig. 9.8. Simple attraction

effect that the context state will move towards the closest TCAD-stable state, as illustrated in Fig. 9.8.

Although this approach seems appealing, the implementation does not produce good results and it is easy to see why. When the context state is close to a TCAD-stable state, say A, then the internal dynamics will force the context state towards A so that it will get closer. Without any environment-driven input, the state would move further towards its attractor.

Such a dynamics produces two side-effects. The first is a sharpening of the percept by the attraction. A move to a TCAD-stable state implies that the meaning becomes more clear. This is desirable, because this is what can be expected from an interpretation. But the second effect, however, is a delay that is caused by the attraction. If the context state is close to A, say, but the environment drives the context state from A to B, then the interpretation will follow, but with a certain delay because A will still exert force and attract the context state. This effect is undesirable but it can be suppressed by a parameter that scales the force of attraction. However, this would be at the cost of the first effect and finally one would end with an interpretation path that is identical to the perception path. The conclusion is that a correct interpretation can never be found if the interpretation follows the time index of the percept.

The delay caused by attraction is similar to the *hysteresis* effect (Fig. 6.4). Hysteresis occurs at phase-transitions of complex system behavior, in particular, at the points where transitions occur from one stable state to another. The transition is typically delayed by the forces of the attractor but it can be compensated by an interpretation process in the sense that the interpretation of a percept at a certain moment in time involves a reconsideration of past interpretations – going back to a certain time in the past – in the light of new information.

A more useful metaphor is perhaps that of an *elastic snail-like* moving object, whose position in the state space is described by the states contained in the working memory (Fig. 9.1). The head follows the musical present and the tail corresponds to a time-limited past. The trajectory followed by the *head* of the snail is described by a TCAD-recognition analysis. The tail, however, corresponds to a working memory which records the adapted states. The states are adapted according to their position to the TCAD-stable states.

The position of the head is important because it partly drives the tail. When the head changes from one attractor to another, the states of the tail are adapted as well. But there is a competition because the tail itself is susceptible to forces of attraction. So it may happen that a part of the tail remains near one attractor, while the head and another part are near to a different attractor. That is what is meant by "elasticity" : the snail may be influenced by the forces of different attractors.

9.6 The TCAD Model

This section expresses the metaphor of the "elastic snail" as a mathematical model. The concepts are first defined and then used to express the dynamics introduced.

9.6.1 Definitions

The following concepts are defined in the TCAD model:

1. A tone context image at t is a state defined as a vector
$$P(t) = \langle p_n \rangle_{n=1,\dots,N}\,, \tag{9.1}$$
where N is the dimensionality of the context image ($N = 56$ in VAM).
2. A tone center image is a TCAD-stable state defined as a vector
$$T^{k\ (k=1,\dots,24)} = \langle c_n^k \rangle_{n=1,\dots,N}\,, \tag{9.2}$$
where k corresponds to a label such that T^1=**C**, T^2=**C♯**, ...,T^{11}=**B**, T^{12}=**c**, ..., T^{24}=**b**.
3. A semantic image at t is a state defined as a vector
$$I(t) = \langle cor\,(P(t), T^k) \rangle_{k=1,\dots,24}\,, \tag{9.3}$$
where the function $cor()$ computes the distance between $P(t)$ and T^k in terms of a correlation coefficient. $I(t)$ is a vector in a 24-dimensional space spanned by the TCAD-stable states. I can be considered a function that maps elements of a N-dimensional embedding space onto elements of a 24-dimensional space: $I : R^N \to R^{24}$.
4. The P-states are stored in a buffer where they can be adapted. Each state in the buffer follows a trajectory in the state space and has an offset to a percept state – that is the most recent encountered P-state. The buffer acts like a shift-register and the states can therefore be referred to by means of a double index – one for the trajectory (t) and one for the offset with respect to the percept state (τ). The double index is illustrated in Fig. 9.9. The horizontal lines correspond to the offset, the vertical lines to the trajectory. The upper horizontal line contains the percept states.

The notation $P(t,\tau)$ thus means that in the buffer at observation time t, P has offset of τ time steps from the percept state. The trajectory of $P(t,\tau)$ started at $P(t-\tau,0)$. By definition, the most recent state is called $P(t,0)$ or $P(t)$. Examples: $P(10,9)$ is a state whose position is observed at time 10 and whose trajectory started at time 1. $P(t-5,9)$ is a state whose trajectory started at $t-14$, but the state is observed at time $t-5$. A short-term buffer (working memory) at t is an array of vectors defined as

$$\Pi(t,0) = \Pi(t) = \langle P(t,\tau)\rangle_{\tau=0,\ldots,L-1}, \tag{9.4}$$

where τ is the offset relative to the percept state at t and L is the total length of the buffer. $\Pi(t,0)$ describes the snail-like object in the N-dimensional embedding space. $P(t,0)$ is the head and the $P(t,\tau)$ is the tail (for $0 < \tau < L$). In Fig. 9.9, $\Pi(t)$ contains the states along the diagonal line. From the above definitions, it follows that if $\tau > L-1$, the state is no longer contained in the buffer. At that moment, it becomes impossible to follow the trajectory any longer. This state is out of the viewpoint of interpretation.

5. The trajectory of a P-state is described by a corresponding I-state, which reflects the schema response. In general, for each buffer Π, there is a buffer Υ which contains the TCAD-stable state responses. These responses drive

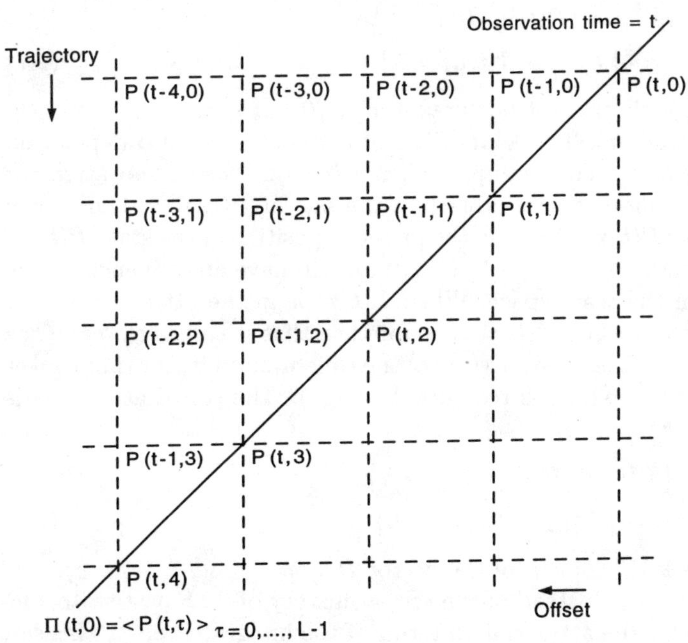

Fig. 9.9. The double index system

the adaptation of the P-states and play a very important role in the TCAD dynamics. A TCAD-response buffer is thus defined as an array of vectors:

$$\Upsilon(t,0) = \Upsilon(t) = \langle I(t,\tau)\rangle_{\tau=0,\ldots,L-1}, \tag{9.5}$$

where $I(t,\tau)$ is defined in the double indexing system as

$$I(t,\tau) = \langle\ cor\left(P(t,\tau),T^k\right)\ \rangle_{k=0,\ldots,24}, \tag{9.6}$$

For example, $I(4,3)$ is the TCAD-response to $P(4,3)$.

9.6.2 Dynamics

During perception, the P-states are attracted by the TCAD-stable states along the information contained in the I-states. In the current version of TCAD, adaptation depends on the following factors :

1. The $P(t,0)$-attractors. When $P(t,0)$ is close to an attractor T, then T will attract not only $P(t,0)$ but to some extent also all other states $P(t,\tau)$ – as $P(t,0)$ partly drives the buffer $\Pi(t,0)$. As a result, the P-states come closer to T and this effect is again reflected in $\Upsilon(t,0)$. In other words: when a state $P(t,0)$ is within a distance h of a stable state T^k, then T^k will attract $P(t,0)$. The distance h can be such that $P(t,0)$ is in the attractor field of multiple stable states. This set is labeled $A(t,0)$:

$$\begin{aligned} A(t,0) &= \left\{T^k|cor\left(P(t,0),T^k\right) > h\right\} \\ &= \left\{T^k|i_k \in I(t,0) > h\right\}, \end{aligned} \tag{9.7}$$

   where i_k is the k-th element of the vector $I(t,0)$.
2. The $P(t,\tau)$-attractors. The adaptation of a P-state will also depend on its proper trajectory. For example, when $P(t,\tau)$ is close to an attractor T – possibly a different attractor than the one of $P(t,0)$ – then T will attract the state $P(t,\tau)$. As such, the previous position of the state $P(t,\tau)$, given by the state $P(t-1,\tau-1)$ (Fig. 9.9), will have an influence on the new position in the state space. When $P(t,\tau)$ is in the attractor field of multiple attractors, then $P(t,\tau)$ is adapted so that it comes closer to the $P(t,\tau)$-attractors. Each state $P(t,\tau)$ of the time-limited buffer thus evokes a schema response, which is recorded in $I(t,\tau)$. The set of attractors is defined by $A(t,\tau)$:

$$\begin{aligned} A(t,\tau) &= \left\{T^k|cor\left(P(t,\tau),T^k\right) > h\right\} \\ &= \left\{T^k|i_k \in I(t,\tau) > h\right\}, \end{aligned} \tag{9.8}$$

   where i_k is the k-th element of the vector $I(t,\tau)$.
3. The Integrated Past. Instead of the proper history of each past state, one may also consider the attraction in terms of the integrated past. This set, called $A(t,\sum)$, can be defined as:

$$A\left(t,\sum\right) = \left\{T^k|cor\left(\frac{1}{L-1}\sum_{\tau=1}^{L-1}P(t,\tau),T^k\right) > h\right\}$$
$$= \left\{T^k|i_k \in I(t,\sum) > h\right\}. \tag{9.9}$$

This equation accounts for the attraction of the whole buffer. It may be used as an alternative or additional constraint to the individual sets of attractors. Another constraint could be the decreasing influence of $P(t,0)$ in function of increasing τ but at present, this constraint is not taken into account.

Obviously, the A-sets introduce a factor of competition. They prevent the interpretation from becoming too dependent on the current percept state $P(t,0)$ and its associated attractor set $A(t,0)$. The competition accounts for the "elasticity" but it has the property that when a tone center was easily recognized in the past, its interpretation will be more difficult to change, even in the light of new evidence. On the other hand, when an object was ambiguous in the past, its interpretation will be easier to change in the light of new evidence.

Given this background, it is possible to formulate the adaptation rule. The adaptation of the n-th element of a state (vector) $P(t,\tau)$ in the light of the new encountered state $P(t,0)$ is:

$$p\,(t,\tau)_n = p\,(t-1,\tau-1)_n +$$
$$\alpha \sum_{k\in A(t,0)} c_n^k cor\left(P(t,0),T^k\right) +$$
$$\beta \sum_{k\in A(t-1,\tau-1)} c_n^k cor\left(P(t-1,\tau-1),T^k\right) +$$
$$\gamma \sum_{k\in A\left(t-1,\sum\right)} c_n^k cor\left(P(t-1,\tau-1),T^k\right). \tag{9.10}$$

The equation is applied to all elements p_n of P and uses the elements c_n^k of T^k. The summations run over all T^k that satisfy the above conditions for (9.7–9) respectively. The parameters α, β, and γ are scaling factors that define the rate of adaptation.

Equation (9.10) says that the adaptation of the n-th element of the state $P(t,\tau)$ is based on the n-th element of the previous state $P(t-1,\tau-1)$ plus changes effectuated by

- attractors of $P(t,0)$,
- attractors of $P(t-1,\tau-1)$,
- attractors of $P(t-1,\sum)$.

The adaptation depends on the distance (correlation) between P and T. When both states are close enough to each other then P will move a little bit in the direction of T. This depends on whether T is a member of the A-set. Obviously, (9.10) affects only the states contained in the buffer $\Pi(t,0)$. $P(t,0)$, itself, is not adapted.

To allow a better concurrence between the members of the A-sets, the correlation values are normalized with respect to h.

9.7 TCAD – At Work

One of the problems of a TCAD-interpretation concerns the representation of the output. For each time point in the running time, there is a memory buffer which updates the states, so that, normally, the output should be conceived along four dimensions: (i) tone center, (ii) similarity to tone center, (iii) running time frame, (iv) and offset. The dynamics could be visualized with the aid of a movie but for practical reasons this is here reduced to an output at fixed intervals of 0.1 s. In particular, the states are shown just before they leave the memory buffer. The adaptation time is taken to be 3 s.

Figure 9.10 shows the TCAD-interpretation analysis of *Arabesque No. 1* based on the following parameter settings: $\alpha = 1$, $\beta = 0.5$, $\gamma = 0$, $h = 0.73$. The backward adaptation effect is illustrated at the beginning. The single note **MI** is now mainly interpreted to be in **C**. The reference to **F**, however, has remained – although the difference with **C** is very small (the difference

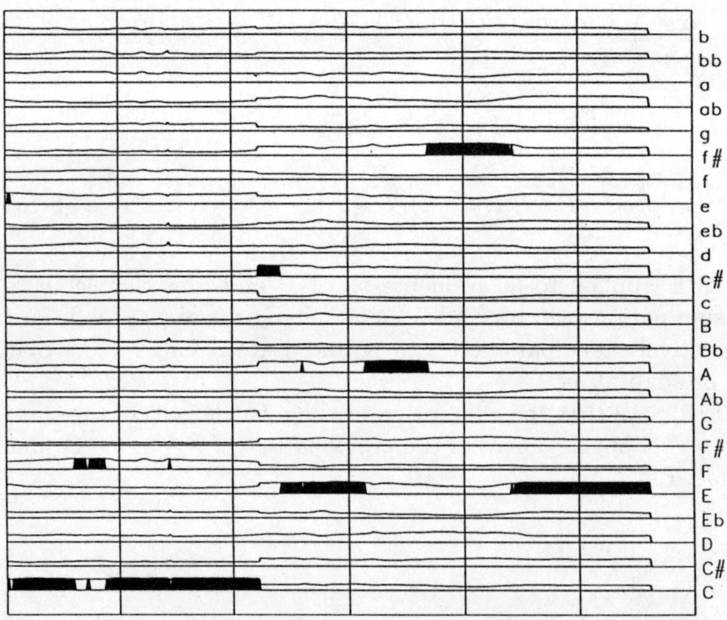

Fig. 9.10. Semantic images based on TCAD-interpretation ($\gamma = 0$)

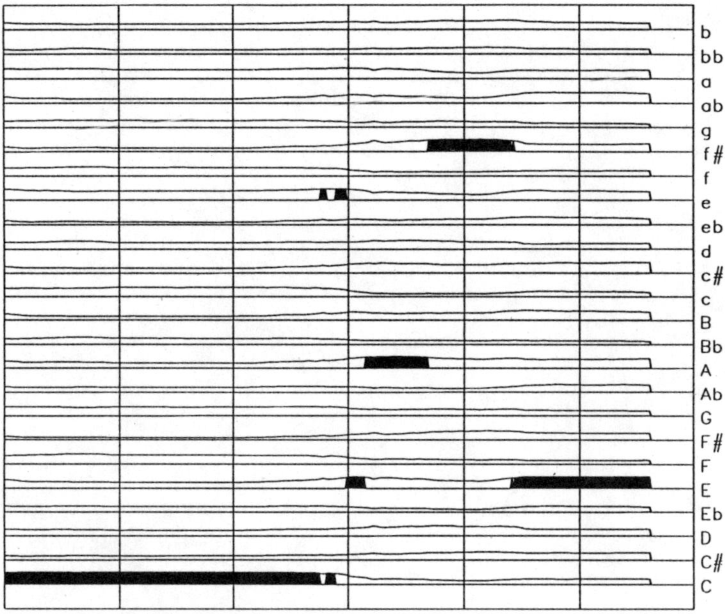

Fig. 9.11. Semantic images based on TCAD-interpretation ($\gamma = 0.25$)

in correlation coefficient is about 0.02). In the second section, the reference to **d** has gone. In the third section, there is now a clear demarcation between the first part (in **C**) and the second part. This effect is explained by the so-called "elasticity" of the "snail". In addition, the "hesitations" in the second part of section 3 (Fig. 9.5) are replaced by a pronounced decision in favour of **E**, although also **A** has a high score. Delay effects, which were due to the integration, are now resolved, and a demarcation (the jump from **C** to **c♯**) is clearly visible. Sections four, five and six confirm the above observations.

The introduction of the integrator has thus far been less fruitful. Figure 9.11 shows a TCAD-interpretation analysis with $\gamma = 0.25$ (all other parameters remain the same). As a result of integration, the reference to **F** in the first section disappears, but integration seems to have a negative effect on the recognition of the modulation (section 3). Further experiments with parameters may point out that the influence of the integrated past should be considered less important.

9.8 Conclusion

In this chapter, a model of associative self-organization has been developed. The model, called TCAD, is a framework for the study of tone center perception. Its behavior is described in terms of an internal attraction dynamics which is driven by a context-sensitive preprocessor.

10. Evaluation of the Tone Center Recognition Model

After a short overview of other models for tone center recognition, this chapter evaluates the tone center recognition model by applying it to musical examples of Chopin, Brahms, and Bartók. The examples belong to the tonal repertoire and have been selected as an illustration of the power and limits of the model.

10.1 Overview of Other Models

In the recent past, several models for tone center perception have been proposed. They differ from the present one in that:

- the distinction between data-driven long-term dynamics and schema-driven short-term dynamics is not made,
- the role of an auditory model is often neglected,
- there is no such thing as schema-based "interpretation" in terms of a reconsidered past.

The models can be classified as *symbolic* or *non-symbolic*. In the latter category, a distinction can be made between *spreading activation*, *supervised* and *self-organizing* models.

The symbol-based models go back to the work of *Longuet-Higgins* and *Steedman* [10.8-10], and *Holtzman* [10.7]. These models start from a symbolic representation of music based on a score and they often include harmonic analysis as well. The analysis is implemented by means of a rule system. Models, such as *Maxwell* [10.11], are able to give an analysis in terms of harmonic functions but the explanatory value is restricted because there is no justification in terms of an underlying auditory system.

Bharucha [10,1] has developed a connectionist model for the explanation of "perceptual facilitation". Facilitation is a measurable effect of musical expectation processes that can be extracted in our model by considering the distance of the tone context to the tone centers. (This information is contained in the interpretation vectors.) Bharucha's model is based on spreading activation. The representation is local (there is a one-one mapping from concepts – notes, chords and keys – to the nodes of the network), and the connections are predefined. The solutions are found by spreading of activation from tone units to chord and key units until a state of equilibrium is achieved. There is no underlying auditory model, hence it is not clear how such a model can develop by data-driven self-organization.

The perceptron networks (based on the backpropagation learning algorithm), extended with a feedback, accumulator and forgetting function have been used to store sequences of patterns [10.2, 13]. By feedback it is possible to accumulate information of the past and a forgetting function limits the accumulation over time. The method is related to the integration technique that we use for the tone context images. In the model of *Bharucha* and *Todd* [10.2], however, there are no compelling forces to learn the sequences, so that, in principle, it is possible to teach the network any chord series. The output will always reflect the probability distribution of the series learned. In other words, there is nothing in the network by which the relations between chords follow from the intrinsic properties of acoustic and psychoacoustic nature.

A final category of networks is based on self-organization. Recently, *Gjerdingen* [10.5, 6] has developed a model to learn syntactically significant temporal patterns of chords based on the ART architecture for neural networks [10.3]. It has a dynamic short-term memory with a retention function and a categorizing network that categorizes the patterns on the basis of their similarity to one another. The model is perhaps closest to our model in that it involves a short-term dynamics as well as a long-term dynamics. As in the previous models, however, it is not clear how to connect the model with an auditory model and there is no "backtracking" interpretation mechanism involved.

In the following sections, we discuss a procedure for the evaluation of the TCAD model and give some concrete examples of musical pieces analysed by the model.

10.2 TCAD-Based Tone Center Analysis

The TCAD-analysis is based on the procedures discussed in previous chapters. Summarized briefly, the following steps have been taken:

1. The audio signal (CD-quality) of a musical excerpt is sampled and converted to 20000 sa/s in order to fit with the sampling rate of the auditory model.
2. The signal is processed with VAM, using the parameter settings described before[1].
3. The VAM completion images are integrated into context images ($w = 3$ s) and normalized according to the Euclidian Norm.
4. The TCAD-analysis is either passive or active. The first is called a *TCAD-recognition* analysis, the second is called a *TCAD-interpretation* analysis.

---

[1] See p. 57.

The parameters of the TCAD-interpretation analysis (Sect. 9.6.2) are: $h = 0.73$, $\alpha = 1$, $\beta = 0.5$, and $\gamma = 0$.[2]

5. Since VAM produces a completion image every 10 ms, this is also the rate at which TCAD works. The semantic images, however, are recorded at a lower sampling rate of 10 sa/s (a time interval of 100 ms). These images convey the position of a state within the framework of the tone center images. In the TCAD-interpretation analysis, the states (that fall within an attraction field) are adapted. The maximum time of adaptation is 3 s, and the snapshots, taken every 0.1 s, pertain to those adapted states.

In terms of the schema theory, the semantic images record the tension/relaxation between the unstable context images and the learned stable tone center images of the schema.

10.3 The Evaluation Method

The semantic images are examined by a musicologist who marks every computer output as (+), (-), or (o) according to the following criteria:

- a plus-sign means that the output corresponds with the musicologist's analysis of the score,
- a minus-sign means that the analysis does not correspond with the musicologist's analysis of the score,
- a zero-sign means that the computer output is one acceptable interpretation of the score.

It is important to note that the present evaluation procedure relies on music theory and musical intuition, and does not take into account any experimental perceptual analysis. The evaluation should therefore take into account possible constraints of the perception-based computer output. Two practical problems had to be solved:

1. The vector representations which stand for semantic images are hard to manipulate in a musicological analysis. The semantic image was therefore further reduced to the four highest correlation values – in addition to the corresponding labels for the tone centers. The four highest values suffice to give an idea of the schema response to the tone context image because the underlying geometry of the tone center map is two-dimensional.
2. The musicological analysis is based on the score, while the computer results are based on the audio signal. Therefore, the two sources must be synchronized – for example, by using a sound editor program to set timemarks on the score. In the examples that follow, synchronization marks have been drawn on the score at intervals of 1 s. One such interval thus corresponds with 10 semantic images.

[2] Test with γ show that this parameter is perhaps not very useful.

In addition to these practical problems, the musicological analysis should also take into account the possible ambiguities of tone center recognition. In the TCAD-analysis, this is reflected in the distributed representations of the semantic images.

10.4 Bartók – *Through the Keys*

The piece *Through the Keys* is based on a CD-recording played by D. Ránki (piano).[3] The two voice piano piece is in two parts. Part A (Fig. 10.1) is written in unisone (parallel octaves), while Part B (Fig. 10.2) is in two-voice harmony. In both parts, the tone center modulations are induced by the melody but in the first part, there is no support for any harmonic structure in unisone, while in the second part, there is only very little harmonic support. Given some uncommon modulations, tone center determination is at points ambiguous. This is not an easy piece to start with, but it was basically chosen to show some of the expected shortcomings of the model.

The main keys to which the melodies allude (in both parts) are given by the following tone centers: **D** (measure 1), **e** or **G** (m. 3), **A** (m. 5), **f♯** (m. 7), **E** or **A** (m. 8), **A♭** (m. 15), **B♭** (m. 17), **g** (m. 20), **C** (m. 22), **D** (m. 23).

10.4.1 Analysis

Figure 10.3 shows the tone completion images of measures 13–16 of Part A. The duration of the excerpt is from 11.32 s to 15.04 s. The marks on the score (at intervals of 1 s) help to synchronize the musical notation with the time flow of the computer analysis. The onsets, as well as frequencies, are clearly represented in the completion images.

The tone context images are shown in Fig. 10.4. A list of the reduced semantic images (TCAD-recognition analysis) of the short excerpt is shown in Fig. 10.5. The first column contains the marks of the evaluation, the second column a count of the samples. The numbers should be divided by 10 to obtain the time in seconds. The next four fields contain the highest values of the semantic image, with a symbolic indication of the tone center.

To get a general overview of the TCAD-analysis, graphs are made which show the evolution of the semantic images. The TCAD-recognition analysis is shown in Fig. 10.6 and the TCAD-interpretation analysis is shown in Fig. 10.7. The black colored strips point to the highest values ($> h$) at each time point (horizontal axis). The vertical lines mark sections of 3 s.

10.4.2 Discussion

The musicological evaluation, of which an example is shown in Fig. 10.5, can be summarized in a table, one for Part A and one for Part B (Tables 10.1–2).

Through the Keys
À travers les tonalités
Von Tonart zu Tonart wandernd

Fig. 10.1. *Through the Keys* – Part *A* (B. Bartók) (©Copyright 1940 by Hawkes & Son (London) Ltd.. Definitive corrected edition ©Copyright 1987 by Hawkes & Son (London) Ltd.. Reproduced by permission of Boosey & Hawkes Music Publishers Ltd.)

The TCAD-recognition analysis of Part *A* and *B* shows 39% *correct* answers, 36% *wrong*, and 25% *acceptable*. In the TCAD-interpretation analysis, 46.5% is *correct*, 25%is *wrong*, and 28.5% is *acceptable*. If *correct* answers and *acceptable* answers are taken together, then the scores are 64% and 75%. The

[3] CD: Béla Bartók, Microcosmos – For Children, Teldec 9031-76139-2, 1977/1992.

Fig. 10.2. *Through the Keys* – Part *B* (B. Bartók) (©Copyright 1940 by Hawkes & Son (London) Ltd.. Definitive corrected edition ©Copyright 1987 by Hawkes & Son (London) Ltd.. Reproduced by permission of Boosey & Hawkes Music Publishers Ltd.)

improvement of 11% in favour of the TCAD-interpretation analysis is on the account of Part *A*. In Part *B*, the scores are about equally well (or bad) for both TCAD-recognition and TCAD-interpretation.

It is important to note that the correlation values expressed in the reduced semantic image are not very high. In Part *A* (TCAD-recognition analysis), only 48% of the semantic images contain at least one correlation value higher than $h = 0.73$ – the threshold of attraction. Although the evaluation takes

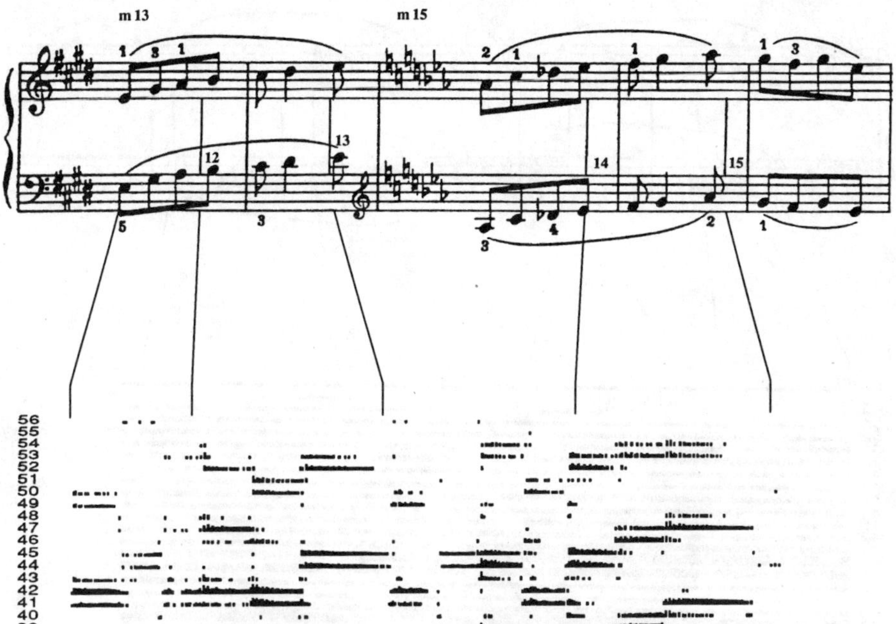

Fig. 10.3. Tone completion images of 11.32–15.04 s (measures 13–16) of *Through the Keys* – Part *A*

into account all semantic images, this low percentage of high correlation values suggest that the recognition capability is perhaps not very reliable.

Figures 10.3–7 illustrate the TCAD-analysis and its evaluation in more detail. At measure 15, the key of **E** suddenly changes into **A**♭ but the tones at the beginning of measure 15 (**LA**♭**-DO-RE**♭**-MI**♭) might be interpreted

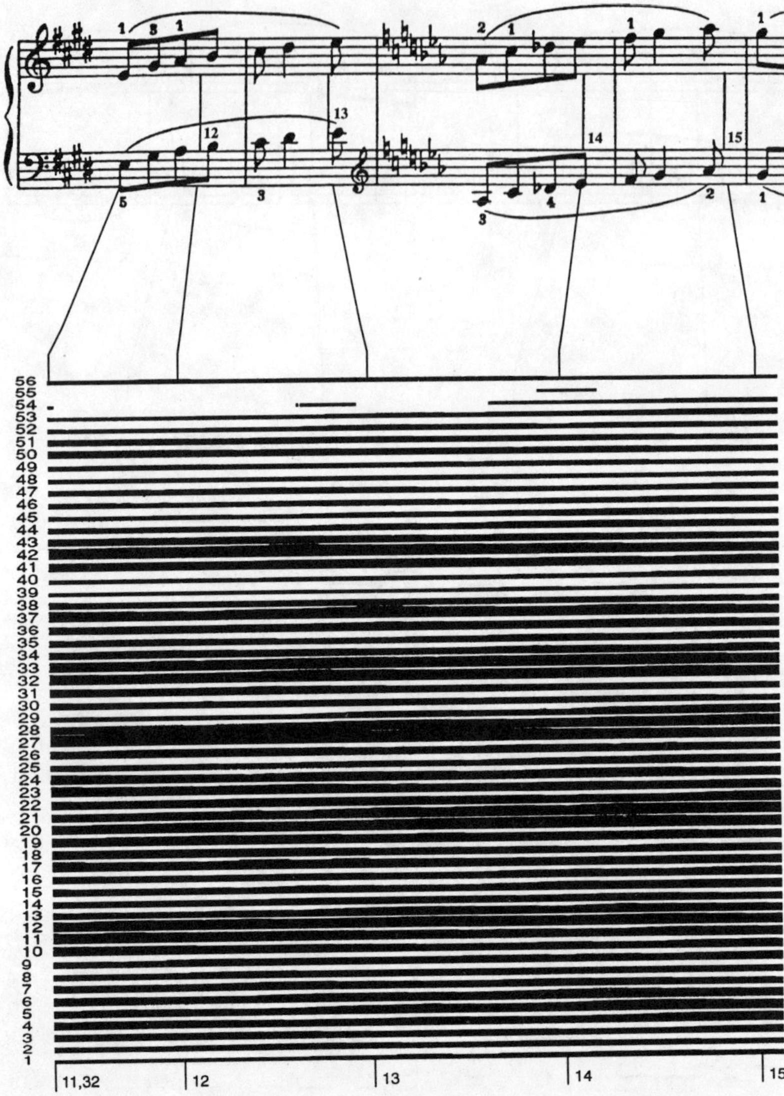

Fig. 10.4. Tone context images of 11.32–15.04 s (measures 13–16) of *Through the Keys* – Part *A*

as belonging to the tone center of **E** (except the **DO**). In fact, this is what TCAD did. The computer interpretation, however, was not accepted by one musicologist who judged the computer output to be wrong. In Fig. 10.5 this is indicated by the marks in the first column starting from line number 133 (=13.3 s). Another musicologist's evaluation was more tolerant and his evaluation showed a correct answer up to 14.0 s (line number 140). In this analysis

```
+   113  E(0.74), e(0.69), f#(0.67), A(0.66)
+   114  E(0.74), e(0.68), f#(0.67), A(0.66)
+   115  E(0.74), e(0.68), f#(0.67), A(0.66)
+   116  E(0.75), e(0.69), A(0.67), B(0.64)
+   117  E(0.75), e(0.70), A(0.66), B(0.62)
+   118  E(0.76), e(0.70), A(0.66), B(0.61)
+   119  E(0.77), e(0.69), A(0.67), c#(0.62)
+   120  E(0.78), e(0.69), A(0.68), c#(0.64)
+   121  E(0.79), A(0.70), e(0.69), c#(0.65)
+   122  E(0.80), A(0.72), e(0.69), c#(0.64)
+   123  E(0.80), e(0.71), A(0.70), c#(0.61)
+   124  E(0.79), e(0.71), A(0.68), B(0.60)
+   125  E(0.78), e(0.70), A(0.68), f#(0.61)
+   126  E(0.77), A(0.69), e(0.68), f#(0.63)
+   127  E(0.76), A(0.70), e(0.66), f#(0.65)
+   128  E(0.78), A(0.70), e(0.67), f#(0.64)
+   129  E(0.79), A(0.69), e(0.68), B(0.65)
+   130  E(0.79), A(0.68), e(0.68), B(0.67)
+   131  E(0.79), B(0.68), e(0.68), A(0.67)
+   132  E(0.79), e(0.69), B(0.68), A(0.67)
-   133  E(0.80), e(0.70), A(0.68), B(0.67)
-   134  E(0.80), e(0.70), A(0.68), B(0.66)
-   135  E(0.82), A(0.69), e(0.69), B(0.66)
-   136  E(0.83), A(0.69), c#(0.68), e(0.68)
-   137  E(0.83), c#(0.72), A(0.70), e(0.69)
-   138  E(0.82), c#(0.73), A(0.70), e(0.70)
-   139  E(0.81), c#(0.75), A(0.71), e(0.68)
-   140  E(0.79), c#(0.76), A(0.71), ab(0.66)
-   141  E(0.78), c#(0.76), A(0.71), ab(0.67)
-   142  E(0.77), c#(0.75), ab(0.70), A(0.68)
-   143  E(0.76), c#(0.73), ab(0.71), A(0.66)
-   144  E(0.76), c#(0.73), ab(0.71), A(0.66)
-   145  E(0.75), c#(0.73), ab(0.70), B(0.66)
-   146  E(0.74), c#(0.72), ab(0.69), B(0.66)
-   147  E(0.74), c#(0.72), ab(0.70), B(0.66)
-   148  E(0.73), c#(0.72), ab(0.71), e(0.66)
-   149  E(0.72), ab(0.71), c#(0.70), e(0.67)
```

Fig. 10.5. Reduced semantic images of *Through the Keys* – Part *A*. The first column contains marks, the second column the line number (which corresponds to tenths of seconds), the fourth column contains four tone center labels and corresponding correlation values

Table 10.1. TCAD-recognition analysis of *Through the Keys*

	Part A	Part B	Total Percentage
correct	35	43	39
acceptable	30	20	25
wrong	35	37	36

(not shown here) the remaining outputs (from 141 to 149) were found to be acceptable and the global evaluation shows a slightly better score. Musicologists indeed might differ in opinion about what is an acceptable answer because the evaluation has its ultimate justification in musical intuition.

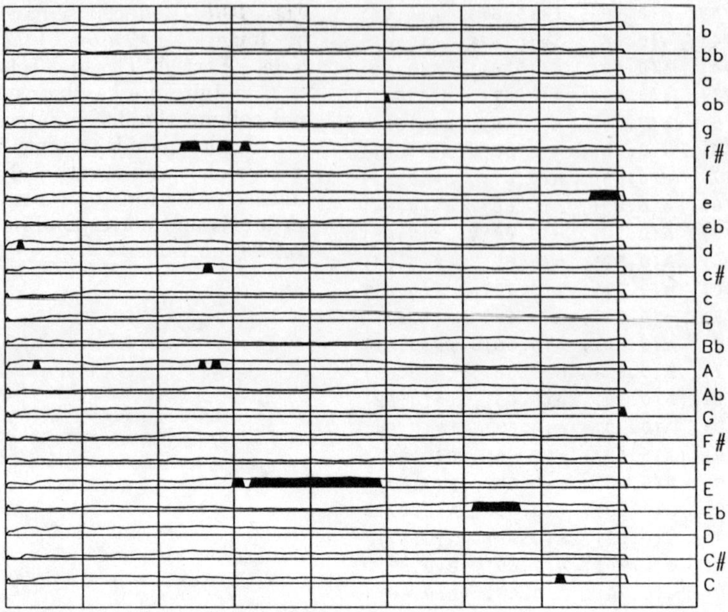

Fig. 10.6. TCAD-recognition analysis (semantic images) of *Through the Keys* – Part *A*

Table 10.2. TCAD-interpretation analysis of *Through the Keys*

	Part A	Part B	Total Percentage
correct	48	45	46.5
acceptable	37	20	28.5
wrong	15	35	25

Taking into account these difficulties, a tolerance range of about 5–10% must be assumed. The evaluation (summarized in Tables 10.1–2) is perhaps severe, and a complementary qualitative analysis is therefore needed. Rather than expressing the evaluation in numeric tables, a careful analysis of the TCAD-behavior has revealed some interesting characteristics. The problems can be divided into four categories:

1. **Melodic Phrase**. In this quasi-monodic piece, cadencial progression depends more on melodic phrase than on harmonic cues. In Part *A*, examples of such cadences are found in measure 5–6 (where the tone center of **A** is established by the notes **RE**, **MI**, and **LA**), measure 18–19 (**B♭**), and measure 23–25 (**D**).
   TCAD does not take into account the effect of phrase and rhythm and this is a major reason for the relatively bad results of this analysis. Phrasal and rhythmical cues indeed provide very important structural information on

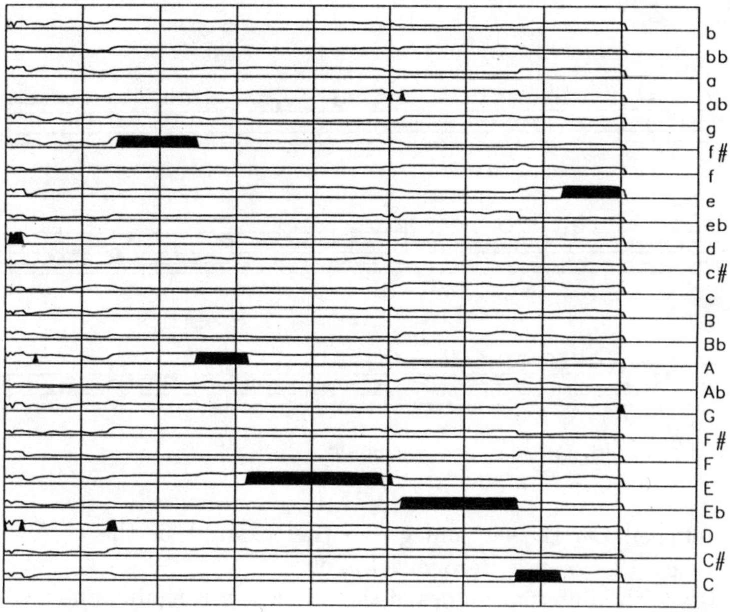

Fig. 10.7. TCAD-interpretation analysis (semantic images) of *Through the Keys* – Part *A*

which the sense of tone center may depend. The cues mark points where a tone center is consolidated or where a transition to a new tone center is prepared. In monodic or quasi-monodic pieces, where the harmonic cues are poor, phrasal cues become more salient.

At present, TCAD processes tone context images which are obtained by a time-integration which is insensitive to phrase and rhythm. Making time-integration sensitive on phrasal cues would improve the analysis (as will be shown in Sect. 10.7).

2. **Leading Tone**. The above examples of Figs. 10.3–7 illustrate that the leading tone may play an important role in tone center perception. Perception of the leading tone, as *Eberlein* [10.4] shows, is an important factor in cadence perception. This factor is again enforced by the melodic character of the piece.
3. **Ambiguity of the Minor Key**. The TCAD-stable states which embody the minor keys reflect the so-called harmonic minor mode. In this mode the seventh is raised (Fig. 10.8). But in music, the old and melodic minor modes are used as well and often they occur together and in mixed forms. In the current implementation of TCAD, both a raised sixth and a minor (normal) seventh degree will affect the tone context image's similarity with the image of the corresponding TCAD-stable state. An example is found in measure 22 of Part B where the prevailing tone center is **a** (melodic mode). TCAD, however, does not recognize it, and is lost

MINOR SCALES:

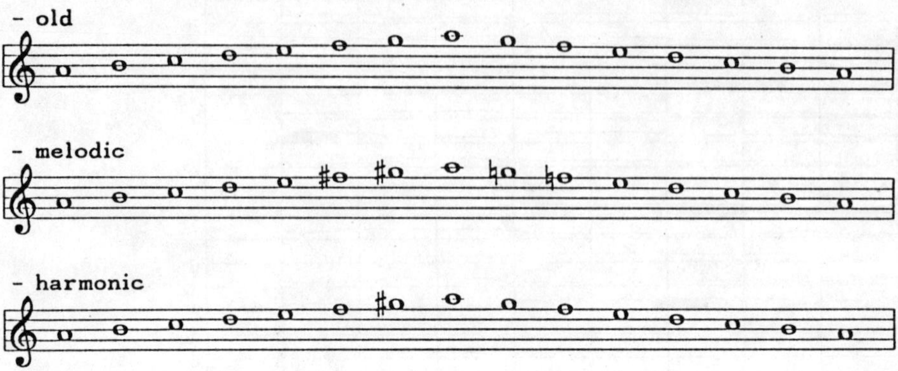

Fig. 10.8. Minor scales: old, melodic, and harmonic

somewhere between **E♭**, **B♭**, **e♭**, and **g**. Another example is the **c♯** (old mode) of measure 12, which is recognized as **E**.
The sensitivity to the different modes of the minor key is again enhanced by the melodic character of the piece. But the results suggest that part of the problem could be solved by adding three different modes of the minor tone centers to the schema rather than one. Since the modes are closely related and often appear together, they should be treated as three modalities of a single perception unit.

4. **The Entanglement of the Parallel Minor-Major**. At several places, TCAD fails to make the distinction between minor and major. This happens for example in measure 15–16 of Part A, where **a♭** is recognized instead of **A♭** – given the natural **DO**. Another example is found in measure 18–19 of Part B, where **f** is recognized instead of **F** – in spite of the natural **LA**. These mistakes are due to a difficult recognition of the third-interval. Although the third is important from a perceptual viewpoint, the interval functionally does not dominate in a harmonic context. The salience of the third is not well represented in the schema and therefore the schema cannot force the interpretation of the tone center when the other tones deviate slightly from the normal pattern. One could try to adapt the schema such as to enhance the relative importance of the third. But this approach could lead to other effects and cause new problems. The entanglement of the parallel minor and major is not independent from the problems discussed in the previous paragraphs.

The evaluation of *Through the Keys* reveals some weakness of the model. The context images, which form the basis of this analysis, are obtained by an integrator which is insensitive to phrase, and the sensitivity to different minor modes is corrupted by the fact that the images which stand for minor tone centers are of the harmonic type (not melodic or old). The TCAD-stable states are *harmony* based entities and the horizontal binding effects of Gestalt perception, of which the leading tone effect is a typical example, are not very well accounted for.

TCAD stresses the harmonic part and neglects the melodic part. This distinction between harmonic and melodic tone center perception has been made by several authors. J. Rameau and H. Riemann seem to favor the harmonic aspect, while E. Kurth stresses the horizontal aspect. The distinction is not in disagreement with recent psychological results [10.4, 12].

10.5 Brahms – *Sextet No. 2*

The second example is taken from the *Sextet No. 2* by J. Brahms in a recording of P. Carmirelli and J. Toth (Violins), P. Naegele and C. Levine (Violas), and F. Arico and D. Reichenberger (Cellos).[4]

The selected fragments are taken from the first part (*Allegro non troppo*) from measure 149 to measure 164 (Fig. 10.9). The piece has been chosen because of its rich harmonic structure. But only in the beginning of the excerpt there is a strong confirmation of the tone center. It is further characterized by tonal instability: cadences are connected to each other by means of alterations and secondary dominants ("Zwischendominante" or "Wechseldominante").

10.5.1 Analysis

The results are shown in Table 10.3. In the TCAD-recognition analysis, there are 53% *correct*, 21% *acceptable*, and 16% *wrong* outputs, while in the TCAD-interpretation analysis, there are 80% *correct*, 11% *acceptable*, and 9% *wrong* outputs. Taking *correct* and *acceptable* outputs together, this gives a total score of 84% for TCAD-recognition and 91% for TCAD-interpretation. In the TCAD-recognition analysis, 75% of the semantic images contain at least one correlation value which is higher than the attractor threshold $h = 0.73$. Taking into account the time needed to build up the context images, this value is rather high. It suggests a good performance of TCAD.

Table 10.3. Analysis of *Sextet No. 2*

	TCAD – recognition	TCAD – interpretation
correct	53	80
acceptable	21	11
wrong	16	9

[4] CD: Brahms, Sony Classical SMK 46249, 1967/1991.

Fig. 10.9. Score excerpt of *Sextet No. 2* (J. Brahms) (measures 149–164 are analyzed)

10.5.2 Discussion

As expected, the results are better than in the Bartók example. The music has a strong harmonic character which TCAD is able to follow. Some of the above problems are recurring, however, – although to a lesser degree. An interesting example is found at the beginning of the Brahms excerpt. In the TCAD-recognition analysis **d** (melodic) is suppressed by **C** and **G**, while in TCAD-interpretation, this problem is solved by the strengthening of the tonal degrees of **d** due to integration. Given a less strong tonal affirmation, the ambiguity would persist.

An example of the harmonic movement is given in Figs. 10.10–11. An overview of the TCAD-recognition analysis and TCAD-interpretation analysis is given in Figs. 10.12–13 (measures 149–164). The black colored strips indicates if the highest correlation value of the semantic image is greater than $h = 0.73$. The vertical lines mark sections of 3 s.

10.6 Chopin – *Prélude No. 20*

This example contains an excerpt from measure 1 to measure 4 of the Chopin *Prélude No. 20* in a recording by V. Perlemuter (piano).[5] Like the Brahms piece, this piece belongs to the romantic (tonal) repertoire. The *Prélude No. 20* has a static harmony with strong cadences. In fact, there is a succession of cadences in different keys.

10.6.1 Analysis

The results are summarized in Table 10.4. In the TCAD-recognition analysis, there are 66% *correct* answers, 24% *acceptable*, and 10% *wrong* outputs. In the TCAD-interpretation analysis, there are 75% *correct* answers, 25% *acceptable*, and no *wrong* outputs. Taking *correct* and *acceptable* together, one obtains a total score of 90% for TCAD-recognition and 100% for TCAD-interpretation. The improvement with the attractor dynamics is again about 10%. In the TCAD-recognition analysis 98% of all semantic images contain at least one correlation value which is higher than $h = 0.73$.

Table 10.4. Analysis of *Prélude No. 20*

	TCAD – recognition	TCAD – interpretation
correct	66	75
acceptable	24	25
wrong	10	0

10.6.2 Discussion

Figures 10.15–16 show the tone completion and tone context images of the first seven seconds of *Prélude No. 20*.

[5] CD Nimbus Records, NIM 5064, 1981.

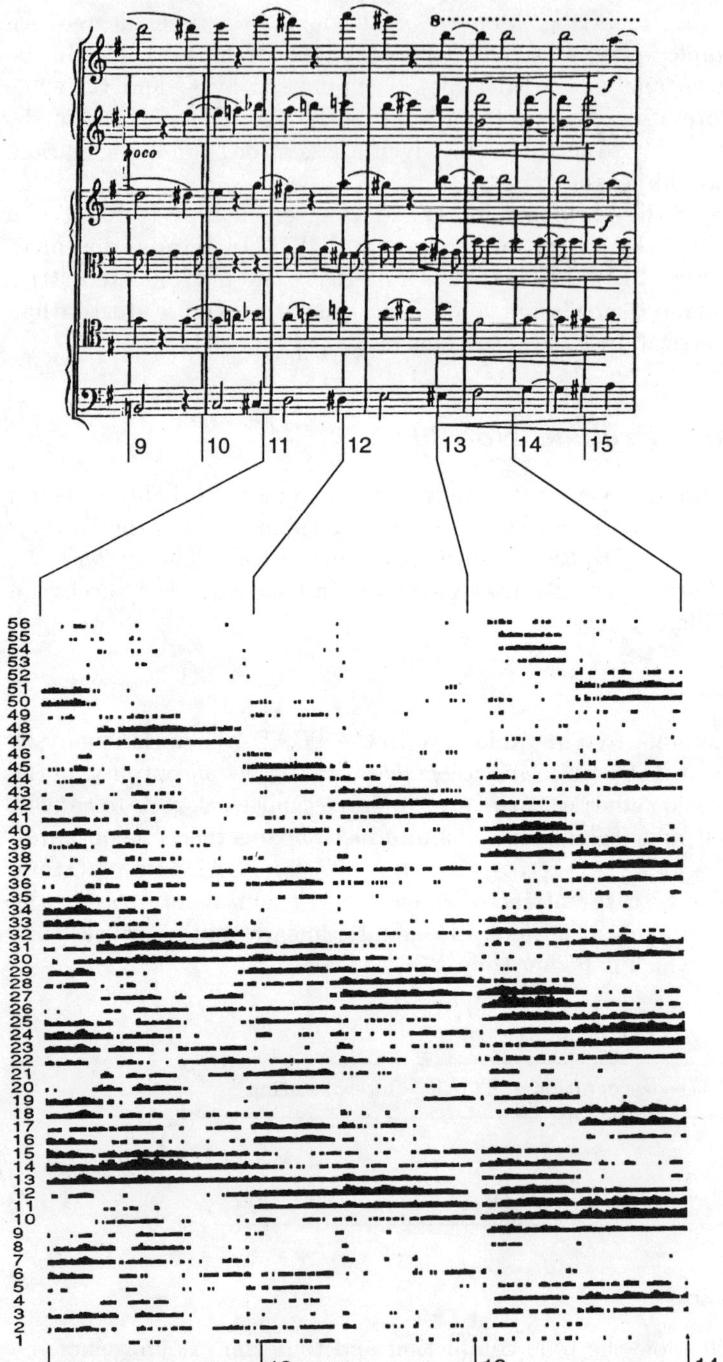

Fig. 10.10. Tone completion images of 11–14 s (measures 158–161) of *Sextet No. 2*

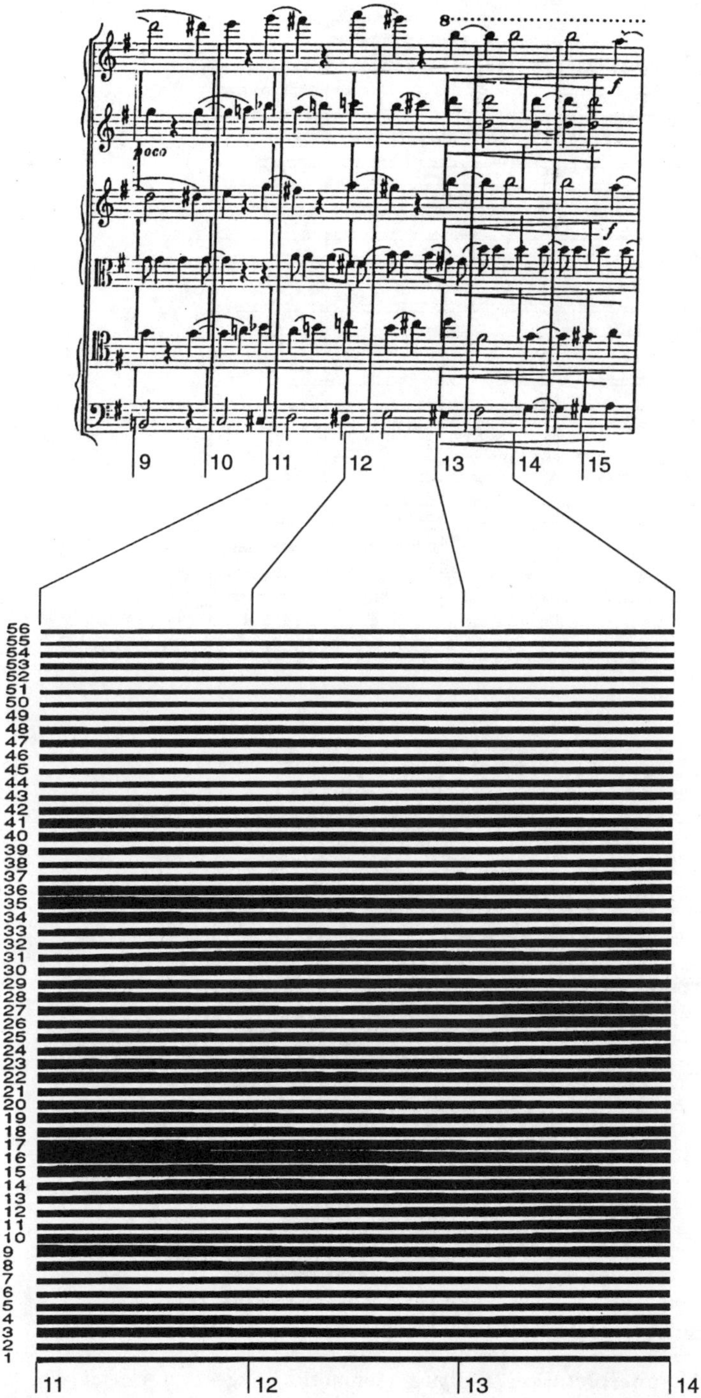

Fig. 10.11. Tone context images of 11–14 s (measures 158–161) of *Sextet No. 2*

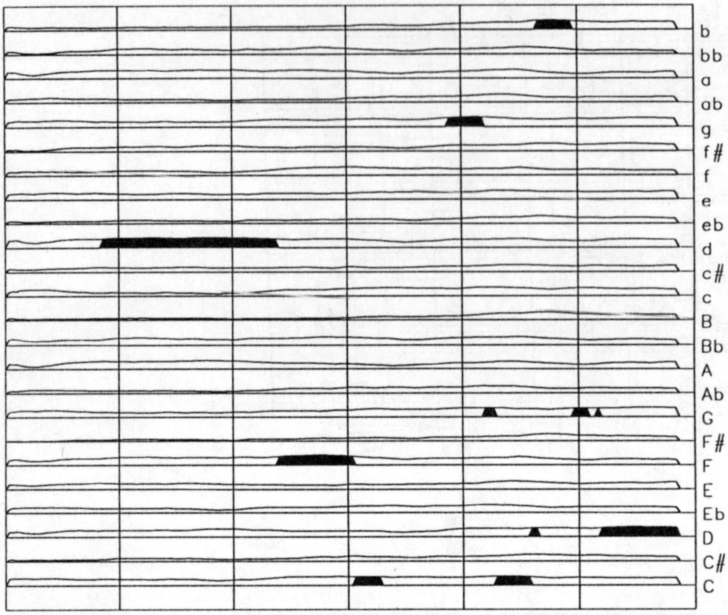

Fig. 10.12. TCAD-recognition analysis (semantic images) of *Sextet No. 2* (measures 149–164)

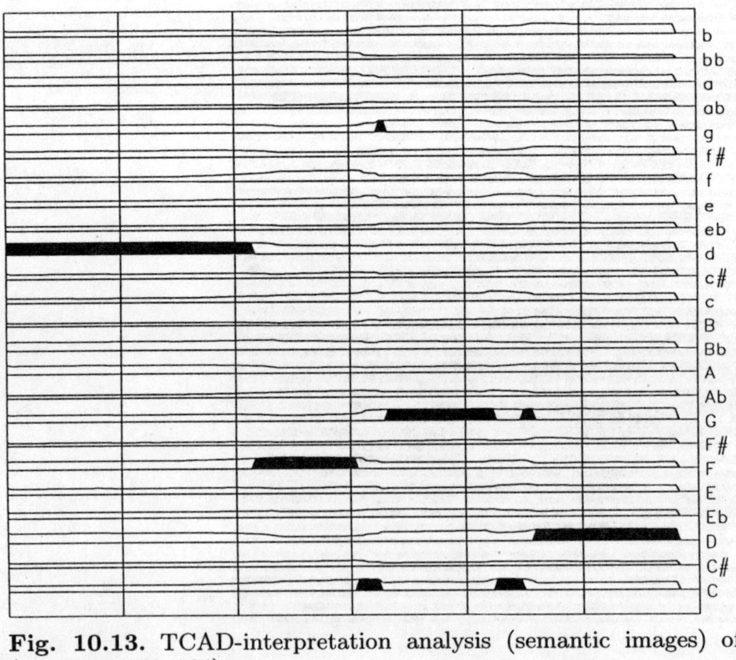

Fig. 10.13. TCAD-interpretation analysis (semantic images) of *Sextet No. 2* (measures 149–164)

Fig. 10.14. Score excerpt (measures 1–6) of *Prélude No. 20* (C. Debussy)

Figures 10.17–18 give an overview of the TCAD-recognition analysis and TCAD-interpretation analysis (measures 1–4). The vertical lines mark sections of 3 s. All cadences are clearly recognized and the only problem seems to be the correct determination of the borderlines. For example, the **A♭M** chord in the beginning of measure 2 is still recognized in the key of **c**, where it is the VIth degree. From a phrasal analysis, it should be the first degree of a cadence in **A♭**. Although this is not a problem of correct recognition, it is again a problem that deals with the segregation of phrasal structures. The entanglement of major and minor is illustrated at the beginning of measure 3. The highest correlation is **c**, although the measure starts with a V-I cadence in **C**. This mistake could probably be solved with a rhythmical grouping module because the preference for the minor is due to the influences from the previous measure.

10.7 The Effect of Phrase – Re-evaluation of *Through the Keys*

In this section, the effect of phrase is examined by a new analysis of the Bartók pieces. The phrases have been artificially defined starting from the score, that is: all notes contained in one legato are taken to be one group. The leaky integrator has an effect within each group but there is no leakage from one group to another. The results are shown in Tables 10.5–6.

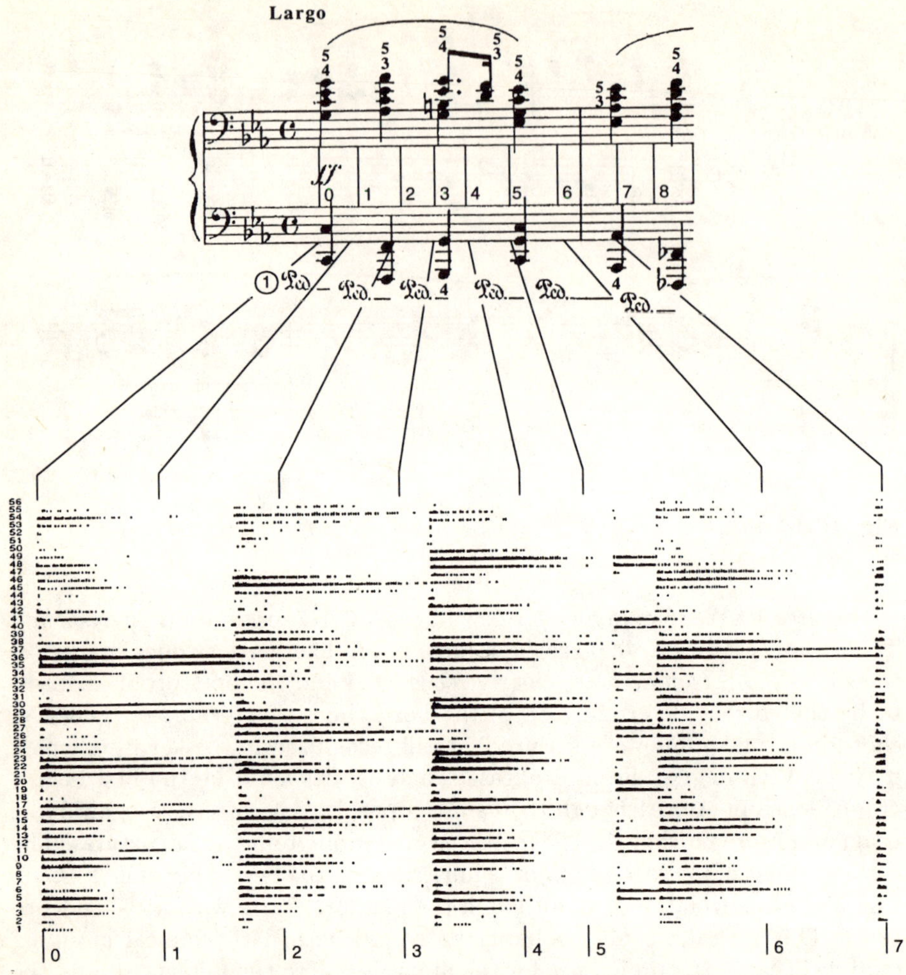

Fig. 10.15. Tone completion images of 0–7 s of *Prélude No. 20*

Table 10.5. TCAD-recognition analysis of *Through the Keys*

	Part A	Part B	Total Percentage
correct	62	64	63
acceptable	31	29	30
wrong	7	7	7

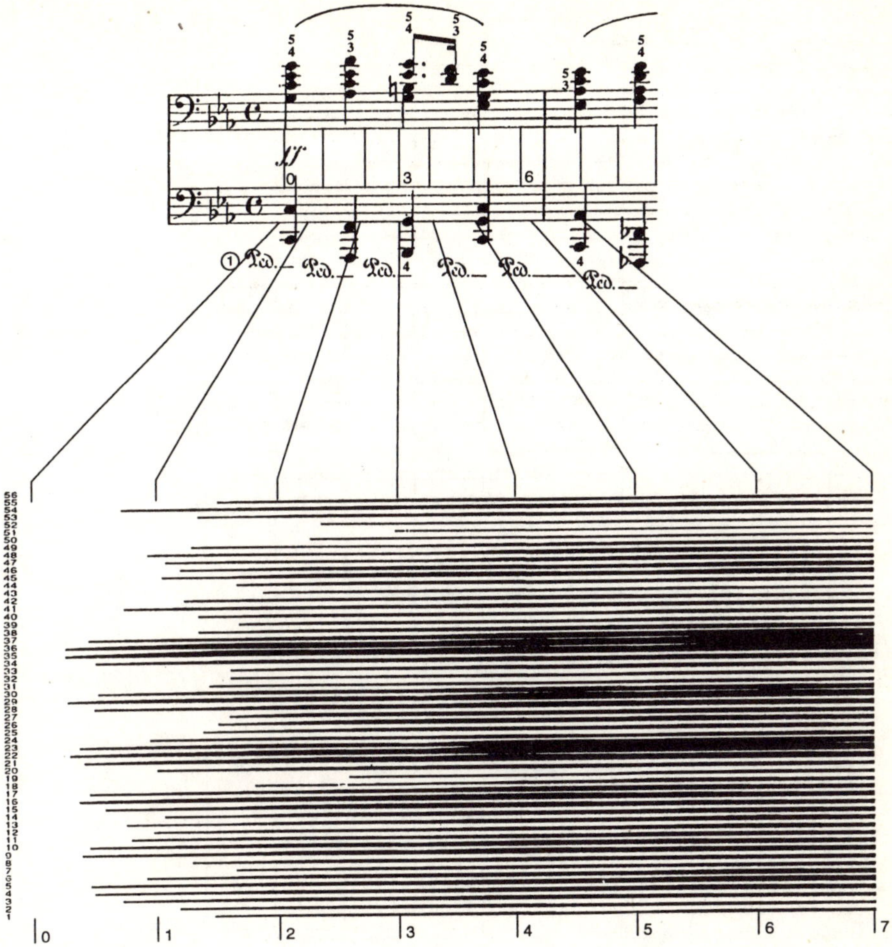

Fig. 10.16. Tone context images of 0–7 s of *Prélude No. 20*

Table 10.6. TCAD-interpretation analysis of *Through the Keys*

	Part A	Part B	Total Percentage
correct	68	73	70.5
acceptable	24	19	21.5
wrong	8	8	8

The results are indeed much better when phrase is taken into account. The TCAD-recognition analysis of Part *A* and *B* shows 63% *correct* answers, 30% *acceptable* and 7% *wrong*. Taking *correct* and *acceptable* together, this gives a score of 93%. The TCAD-interpretation analysis has 70.5% *correct*,

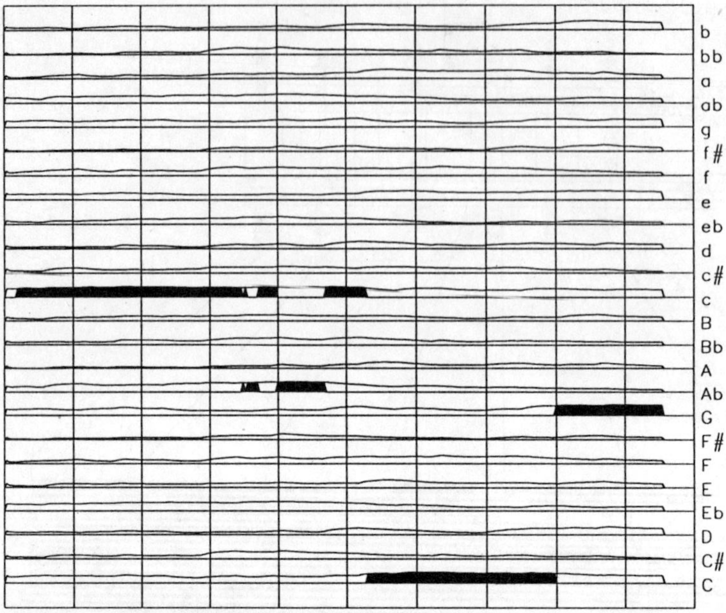

Fig. 10.17. TCAD-recognition analysis (semantic images) of *Prélude No. 20* (measures 1–4)

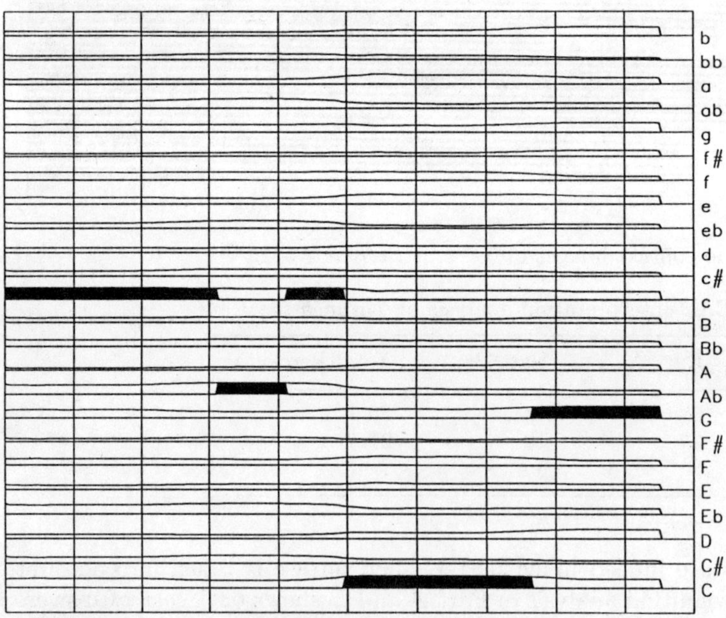

Fig. 10.18. TCAD-interpretation analysis (semantic images) of *Prélude No. 20* (measures 1–4)

21.5% *acceptable*, and 8% *wrong*. Taking *correct* and *acceptable* together, this gives a score of 92%.

There is no big difference between the TCAD-recognition and TCAD-interpretation, an effect which is due to the fact that only 11% of the semantic images in Part A and 16% of the semantic images in Part B have correlation values higher than the threshold of adaptation (> 0.73). This low percentage has of course its effect on the adaptation.

10.8 Conclusion

A TCAD-interpretation analysis of a harmonized piece of music performs about 10% better than a TCAD-recognition analysis. TCAD, however, is good in analyzing harmonic pieces and less good in analyzing melodic pieces. A re-evaluation of TCAD applied to melodic pieces which takes into account the notion of phrase, shows improvements for tone center recognition. This suggests that tone center recognition and rhythmic grouping are indeed influencing each other.

11. Rhythm and Timbre Imagery

This chapter aims to broaden the approach of the previous chapters towards a framework for the study of auditory inter-modular perception. Although many of the subtile interactions between pitch, rhythm and timbre remain beyond the scope of this chapter, an attempt is made to relate these aspects to a general framework of musical imagery.

11.1 Models of Rhythm Perception

Models of rhythmic perception try to explain how the subjective categories of beat and meter emerge in the mind of the listener.

Traditionally, attention has been focused on the problem of meter perception and rhythmic grouping. In the past, music theoretical accounts of this have often been limited, however, to the assumption that note durations correspond to regular inter-onset intervals. In doing so, abstraction is made of the small (and often not so small) irregularities in the inter-onset intervals of played music. In general, the interpretation of meter has been based on backtracking principles (as in rule-based accounts [11.15]) or it relies on techniques of constraint satisfaction (as in connectionist models [11.19]).

Yet, music performers use a number of accents to spice and structure the music they play. Three important categories of accents are:

- *agogical* accents: playing notes shorter or longer,
- *intensity* accents: playing notes louder or softer,
- *rubato* accents: playing accelerando and ritardando.

These accents are part of the performer's palette to give structure and expression to the music. Compared to the pure ratio-division of time and the assumption of constant amplitude, these accents may come across as "deviations". But from a musical point of view, enlargements of durations and increases in amplitude are very natural and essential.

Psychological studies of rhythm indeed confirm that skilled musicians display noticeable deviations from the meter notated in the score. Studies in expressive timing, perception and performance [11.3, 4, 5, 9, 20, 22] show that *context* (tonal as well as rhythmical) plays an important role. Some models rely on contextual factors.

Jones [11.14] relates meter and expressive timing to a theory which associates meter with a reference level that produces the interpretation of the rhythmic pattern from a particular ratio-time perspective. The expressive timing factors introduce non-ratio times that, in Western music, are often related to tonal dynamics.

Computer models aim to give an operational account of rhythmic perception. *Todd* [11.25] relates the effect of accelerando/ritardando to the equations of elementary mechanisms. The concepts of energy and mass are introduced to account for the expressive aspects of rhythm whose ultimate foundation is believed to be based on the vestibular system (not necessarily limited to the cochlear) – where it contributes to the arousal of self-movement. Recently, this author [11.26] has proposed a multi-scale model of rhythmic grouping based on an auditory model. It will be discussed at the end of the section.

Desain and *Honing* [11.7] focus on context-dependent beats, whose function it is to quantize the note durations. Obviously, if the deviations of the onsets from the beat become regular, then a new beat pattern emerges. Since the beat is context-sensitive, it is highly determined by expressive timing factors. The model introduces the beginning of a contextual semantics but is limited to an artificial (and ad hoc) micro-world. Although the approach is interesting, its relevance for auditory systems is far from evident.

In what follows, a model of context-sensitive beat recognition is linked with the auditory model VAM. The model, based on the dynamic paradigm introduced in the previous chapters, considers the beat as a relatively stable (but context-sensitive) perception unit, whose time-base is extracted from the periodicities in the onsets of tones.

11.2 VRAM: A Rhythm Analysis Model

The basic features of VRAM, an auditory model for rhythmic perception, are shown Fig. 11.1. The module consists basically of two parts, one for onset detection (based on the analytical part of VAM), and one for the detection of periodicities in the onset pattern. The onset detector transforms the auditory nerve image into the onset image. The output of the periodicity analysis is called the beat image. The integration of the beat image gives the beat context image.

11.2.1 Detection of Periodicities

Relevant aspects of beat extraction can be illustrated in a simplified environment in which onset-times are given and the nature of the onset pulses is one of the variable parameters whose effect on beat detection can be studied.

In the present analyses, the detection of periodicity is based on autocorrelation [11.2]. When regularities appear in a pattern of onsets, these will

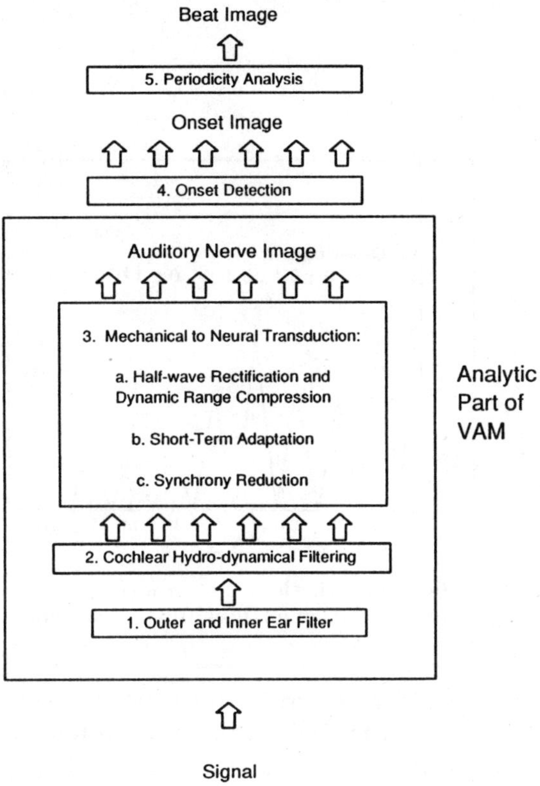

Fig. 11.1. Basic architecture of VRAM: an model for rhythm analysis

be represented by peaks at the time-lags of the periods. Consider the simple rhythmic pattern of Fig. 11.2a. Its representation as a vector is shown in Fig. 11.2b. Such a pattern is called an *onset pattern.* When such patterns are produced by an auditory model, we prefer to call them *onset images.*

The periodicity analysis of the onset pattern is called the *beat pattern* or *beat image* (Fig. 11.2b). The beat marks, generated by the periodicity analysis in the onset pattern, can be interpreted as virtual entities – similar to the virtual patterns obtained by the periodicity analysis of the auditory nerve images (Sect. 5.5).

The sampling rate of the onset pattern defines the resolution of the beat pattern. Fig. 11.2b shows a peak at time-lag 4. Given the resolution of the onset image, this corresponds to a half note (four times the smallest time interval in the onset image). The resolution, however, is not fine, and Fig. 11.2c shows what happens when the resolution of the onset image is doubled. The beat image now displays peaks at regular intervals that correspond to the quarter and half note.

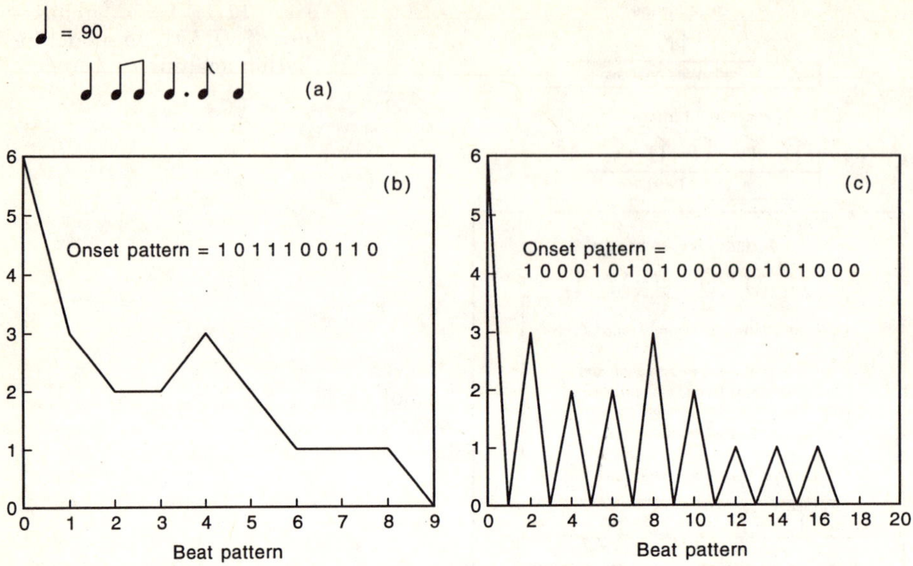

Fig. 11.2. Rhythm pattern and analysis: (a) simple rhythmic pattern, (b) autocorrelation analysis (coarse resolution), (c) autocorrelation analysis (finer resolution)

Obviously, the above periodicity analysis cannot be applied to a whole musical piece. It should be adapted in order to deal with changes in tempo, and the presence of agogical and intensity accents.

1. **Tempo**. Changes in tempo can be accounted for by using a straightforward short-term periodicity analysis, one in which the onset image is analysed in frames at defined frame-intervals. The frame limits the periodicity analysis to a time interval that should be large enough to detect periodicities in the onset pattern. The frame should not be too large, however; otherwise changes in tempo can not be followed. The length of the frame can be based on psychological rhythmic data. Data by *Fraisse* [11.8] suggest beats between 120-1800 ms.
2. **Duration**. The agogical accents introduce small irregularities in the inter-onset intervals. The resolution of the onset image should be fine enough to detect such accents. On the other hand, when onsets are represented by impulses – as in the onset images A and B of Fig. 11.3 – the beat image may not be reliable because the periodicity analysis may miss the periodicity detection by one time-lag. This is illustrated in Fig. 11.4a,b where the onset image is given as a list of inter-onset times (expressed in ms between the onset of the notes.)

   The inter-onset times of A are: 666, 333, 333, 999, 333, 666. The onset pattern is obtained by a transformation of the inter-onset times into ones and zeros. The complete pattern is represented by 100 samples using a sampling time of 33 ms. Fig. 11.4a shows that the beat image is not

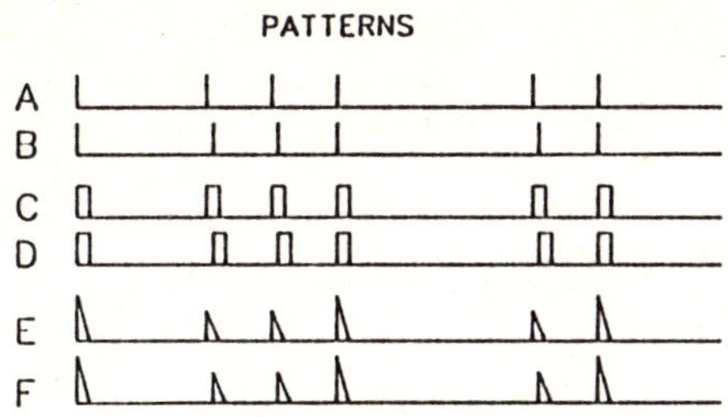

Fig. 11.3. Onset patterns: (A) regular pattern with impulses, (B) irregular pattern with impulses, (C) regular pattern with block-shaped onsets, (D) irregular pattern with block-shaped onsets, (E) regular pattern with ramp-shaped onsets, (F) irregular pattern with ramp-shaped onsets

different from the one in Fig. 11.2b. There are peaks at samples that correspond to $10 * 33 = 330$ ms and $40 * 33 = 1320$ ms.

The inter-onset times of pattern B, shown in Fig. 11.4b, are slightly different: 699, 333, 300, 1042, 300, 666. Some notes are played 33 ms longer, while others are played 33 ms shorter. As a result, the beat at 1320 ms is more prominent, but the difference with the original residue pattern is somewhat exaggerated. A comparison between both beat patterns gives a correlation coefficient of only 0.77.

A better result is obtained by smearing out the onset over more than one unit of the onset pattern. Thus, instead of using impulses, short blocks can be used to mark an onset. This is shown in patterns C and D of Fig. 11.3. C shows the regular pattern and D contains the durational accents. The beat images, shown in Figs. 11.4c,d, have a correlation coefficient of 0.96. (These patterns can be obtained by convolution of the beat-kernel (one block) with the beat patterns of Fig. 11.4a,b).

If an onset differs from the ratio-time of the rhythmic pattern, then it is shifted by one or more sampling intervals, but a smeared onset (rather than an impulse) will guarantee an overlap with onsets that are correct. As such, small deviations can be recovered. If the deviations display a regular pattern, for example by slowing down the tempo, then this effect will be reflected by regular patterns in the frames. At each frame-interval the beat image will mark the beat at larger time-lags.

3. **Amplitude**. Tones that fall on the strong beat are normally played a little bit louder. These intensity accents can be accounted for by the values of the onsets. In Fig. 11.3, the onsets of pattern E (regular) and F (irregular) are represented by ramps of three samples in length. The normal values are: 3 2 1, but a stress on the longer notes was represented by a ramp with the values: 4 2 1. Figure 11.4e shows that the accent on the long notes supports a beat of 1320 ms, even in the regular pattern. In Fig. 11.4f, the peak is enhanced. The correlation coefficient of both images is 0.94.

The above discussion shows that agogical and intensity accents, rather than being "deviations", contribute to the emergence of a beat. The presence of these accents in the musical signal is an important cue for rhythmical grouping, structure, and expression.

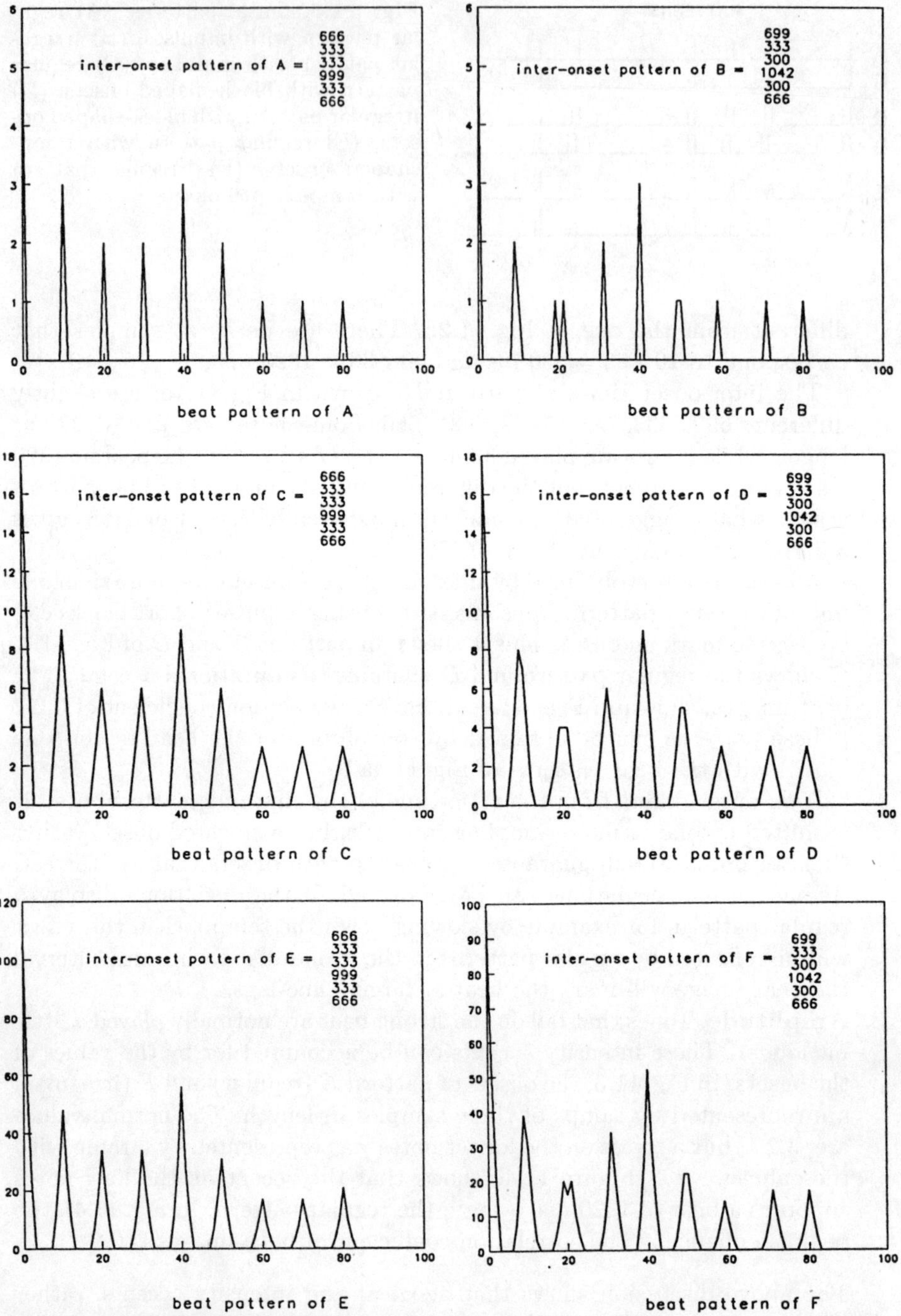

Fig. 11.4. Periodicity analysis of the onset patterns in Fig.11.3: (a-b) analysis of patterns A-B (correlation coefficient=0.77), (c-d) analysis of patterns C-D (corr. coef.=0.96), (e-f) analysis of patterns E-F (corr. coef.=0.94)

11.2.2 VRAM – Analysis

The onset detection part of VRAM (Fig. 11.1) is based on the analytical part of VAM. As discussed in Sect. 5.5.1, the analytical part of VAM transforms the signal into neuronal firing patterns along an array of channels. The channels correspond to auditory nerve fibers whose characteristic frequency is at a distance of one critical zone. The images are called auditory nerve images.

Onset detection in VRAM is based on the fact that certain cells in the cochlear nucleus (the "onset neurons") can extract onsets from the auditory nerve images [11.28]. The present model is based on the assumption that the processing of rhythm is based on a periodicity analysis of the activity in onset-neurons.

- **Onset Detection**. The onset-detector used in VRAM is realized in two steps:
1. The neuronal firing signal in each auditory channel is low-pass filtered (the cut-off frequency is 250 Hz). This allows a down-sampling to 500 sa/s – one onset image every 2 ms.
2. The signal is convolved with a differential onset-kernel, similar to the one used by *Brown* [11.1] and *Mellinger* [11.18]. Another technique for music segregation by means of onset/offset detection has been described by *Smith* [11.24].
- **Periodicity Analysis**. The periodicity analysis is based on the short-term autocorrelation function. This function has been defined in (5.4-6). These are the steps performed:
1. Add up the onset values of over all channels.
2. Perform a short-term autocorrelation analysis every 250 ms using frames of 1600 ms. The parameters are: $K = 2$ s (800 samples), $T = 250$ ms, $\alpha = 0.5$. The parameter α specifies a parabolic attenuation of the autocorrelation pattern at about 600 ms. This value corresponds to the best representation of the natural speed of tapping, or the preferred tempo.
3. Reduce the resolution of the frame K (800 samples) to a frame of 100 units.

11.2.3 VRAM – Examples

Figure 11.5 shows the signal, onset images, and periodicity analysis of the beginning of Chopin's *Prélude no. 7* played by V. Perlemuter.[1] The periodicity analysis is shown in the lower 2/3 of the figure. It is based on the summary onset image which is shown just below the signal representation. The summary onset image adds up all the values of the onset images over all channels. The information contained in these images may be used in grouping analysis and may be related to tone center recognition. *Todd* [11.26] has analyzed the same piece with a multi-scale model for rhythmic grouping. In this model,

[1] CD Nimbus Records, NIM 5064, 1981.

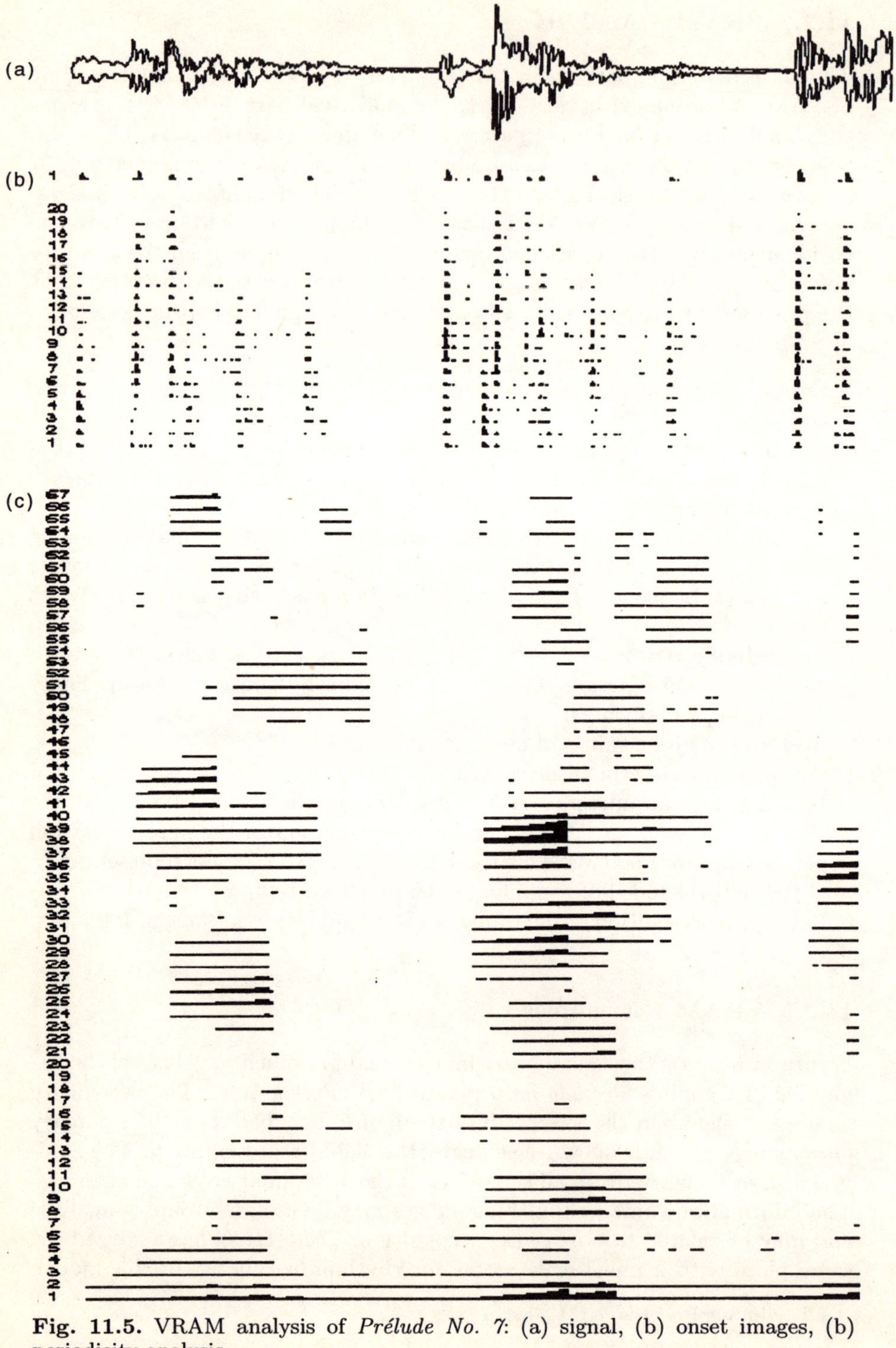

Fig. 11.5. VRAM analysis of *Prélude No. 7*: (a) signal, (b) onset images, (b) periodicity analysis

each channel of the auditory nerve passes a signal to a multi-scale low-pass filter system. This bank of filters is interpreted as a memory system: each filter has a delay or memory which is proportional to the time-constant of the filter. At the other side of this filter system, there is a cell which detects peaks in the responses of the filters. The peaks can be recorded in a time-constant/time graph which is called a *rhythmogram*. Figure 11.6 shows his analysis of Chopin's *Prélude no. 7*.

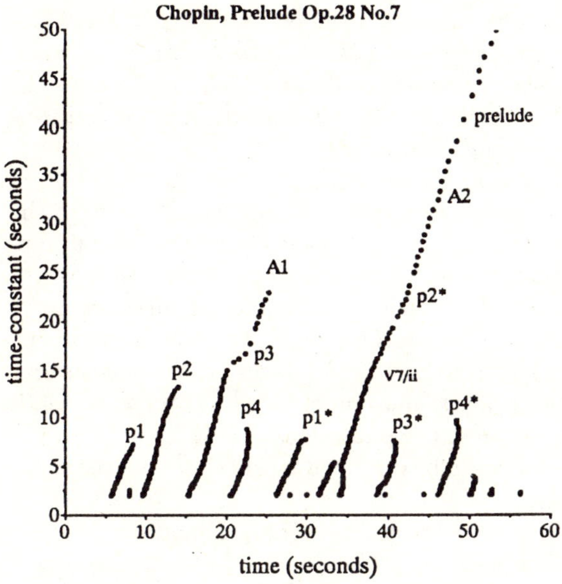

Fig. 11.6. Chopin's *Prélude no. 7*, analysed by the multi-scale model of Todd [11.26]

11.2.4 VRAM – Discussion

The model put forward in this section can be improved and fine-tuned. Yet it illustrates the fact that principles used in pitch recognition can straightforwardly be applied to rhythm analysis as well. In particular the periodicity analysis seems to be quite useful.

Two applications for which the model can be used is quantification and rhythmical grouping. Quantification aims to transform the accents back into a ratio-time divisions. Initial experiments suggest that this can be done by integrating the autocorrelation images into *context images*, and inferring the ratio-time divisions from the number of onsets detected in the interval spanned by the context beat. Rhythmical grouping aims to segment the music in groups or entities. Initial experiments with VRAM suggest that this could be based on jumps detected in the beat patterns (Fig. 11.5).

11.3 The Analysis of Timbre

Cosi et al. [11.6], in a recent paper on timbre analysis, note that timbre is a feature which can hardly be analyzed in physical and mathematical terms due to its dependency on a great number of parameters. In their model they adopt the ear-brain combination as an ontological foundation to musicology. The auditory model is based on *Seneff* [11.23]. The brain model adopted is the Kohonen self-organizing map (SOM) (Sect. 6.3). The research paradigm is therefore similar to the one adopted in our simulations of tone semantics, in particular the studies which have addressed the development of schemata for tone center perception by self-organization. As in tone center recognition, there are a number of psychological studies available which aim to map out the mental representation of timbre [11.10-13]. The output of SOM can thus be compared with these mental representations.

In this paradigm, it is evident that the self-organizing capabilities of SOM owe much to the inherent properties of the spectral images. The role of the auditory model is indeed to extract the features that are relevant for timbre recognition. In the Seneff-model, whose basic design is similar to VAM, a synchrony detector is applied to each channel (standing for an auditory nerve fiber) which implements the known phase-locking property of the nerve fibers. This enhances spectral peaks which are due to resonances in the sound-source and improves the performance in noisy environments. The spectral image is obtained by a short-term integration of the phase-locked activity of the nerve fibers. (In VAM, a spectral vector could be obtained by short-term integration of the activity into each channel.)

Cosi et al. show that SOM is able to classify timbre images into clearly delineated response regions. Timbre images should here be understood as a number of spectral images appended into one single image for each musical instrument. Figure 11.7 gives an example of a classification of musical instruments. The regions identify the timbres on the map.

Earlier classification results on timbre have been obtained by *Toiviainen* [11.27]. Although this model has a preprocessing part based on a Fourier transform, rather than an auditory model, it shows that spectral images, when classified with SOM, produce a trajectory on a map. This trajectory can then be appended in order to obtain the delineated response regions at a second level. Thus far, these studies have worked with a constant pitch and experiments in the classification of timbres of instruments with different pitches are under way.

Timbre recognition, however, is a difficult job, due to the multi-dimensional character of the subject. Its relation to pitch is especially important. The framework, in which images of different kind convey properties which are self-organized at a higher level into a sort of timbre schema seems promising but much more research will be needed to obtain a clearer view on the topic. Perhaps in the future, this approach will make a more detailed account possible of an auditory-based timbre semantics, as an alternative – or extension –

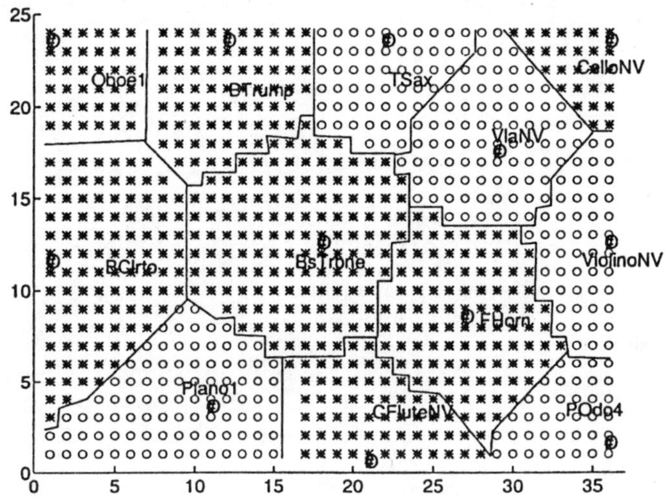

Fig. 11.7. Timbre map based on a SOM architecture. The timbres are divided into regions [11.6]

to the phenomenologically-based timbre semantics introduced by P. Schaeffer in his book on musical objects [11.22].

11.4 Conclusion

Until now, research in cognitive musicology has been dealing with well-defined tasks pertaining to pitch, rhythm, and timbres. A model, in which these modalities are really integrated, does not yet exist [11.16, 17]. Yet, the need for an inter-modular approach is self-evident. Even in a well-defined task such as tone center perception, rhythmical grouping and horizontal binding (for leading tone effects) have to be taken into account.

In the domain of timbre and harmony, research just started. Musicians have been using timbre/harmony relationships in music for a very long time. The theoretical foundations, however, have often been based on distinct categories with the result that the badly understood interrelationships were masked. Musicology should be aware of the fact that pitch, rhythm and timbre are emerging properties of the auditory system and that these properties should be studied from an inter-modular viewpoint.

12. Epistemological Foundations

The previous chapter has broadened the scope of the model to the study of rhythm and timbre. This chapter discusses the foundation of the model in depth and gives an analysis of the basic principles on which a schema theory of music cognition ultimately rests.

12.1 Epistemological Relevance

Tone semantics has been related to schema-based adaptation and control, and it has been argued that this dynamics is ultimately guided by the laws of sound acoustics, gestalt formation, and the culturally-bound stimuli of an environment. Generalized, this ecological perspective involves a physical environment, a neuropsychological system, and a cultural embedding.

In this setup the causal chain from signals to auditory images and schemata is central. The computer-based approach, which means that all of this is modelled in terms of an algorithm, is in many ways *under-determined.* Once an algorithm is implemented, it is easy to find alternatives. Often different architectures can be designed that will do the same job.

The question is therefore: how do computer models relate to the real world? In other words, how can one be sure that the causality simulated by the model is indeed an aspect of the processes that underlie perceptive and cognitive behavior?

This is an epistemological problem and an intricate one. Consider a situation in which two models, A and B, are available and they simulate the same cognitive phenomenon, using very different principles of encoding and processing. Their relevance cannot just be assessed through the ability to predict the correct phenomena because both have an identical predictive value. What other criteria are then needed to decide on the epistemological relevance of these models?

A straightforward criterion, apart from prediction, is that the representational levels of the model should be compatible with neurophysiological data. A computer model indeed has a hierarchical structure with different conceptual levels. Each of the representational, algorithmic and conceptual structures of the model should be relatable to neurophysiological data.

On this basis one could then state that a model has a high degree of E(pistemological)-relevance when its internal structure can be related to neurophysiological data. A high degree of E-relevance then corresponds to a high degree of explanatory power and, ultimately, cognitive musicology should attempt to develop models with a high degree of explanatory power.

In practice, however, it turns out that this reductionist criterion may strand on practical problems. Up to a certain level, the models can rely on neurophysiological data, but very often these data do not provide sufficient information. Models are cues for theory building and they guide empirical research. Rather than being based on empirical data, models are often used to get inspiration for gathering empirical data. The scientific relevance of models cannot therefore be limited to the reductionist criterion.

As known from the philosophy of science, the reductionist criterion (introduced by R. Carnap as a means to evaluate theories) has often been amended, because of the difficulties in practical applications. Some philosophers therefore argue that reductionism should be conceived of within a broader field of *knowledge justification* [12.2, 3]. This view entails that the evaluation of epistemological relevance of a model should be related to a number of *pragmatic factors.* A scientific *context* or research paradigm includes methods, beliefs, the status of the field with respect to other sciences, and the epistemological foundation of the scientific paradigm to which the model is attributed.

When the scientific context changes, then the degree of epistemological relevance may change as well. In other words, the degree of E-relevance of a model is context-sensitive, tentative, and sensitive to changes.

Against this background, this chapter aims to verify the basic principles of the model by means of the reductionist criterion on the one hand, and by means of the context of cognitive science, in particular theories of meaning, on the other hand.

12.2 Neurophysiological Foundations

The representational basis for our schema-based tone center perception model rests on two grounds: evidence for *images* and *schemata.*

12.2.1 Foundations of Images

Auditory models play a fundamental role in the study of music perception as preprocessor devices: they transform the acoustical signal into images in which features of the signal are filtered and enhanced.

The first question is therefore whether the notion of an image, such as defined in the current framework, makes any sense form a neurophysiological point of view. Recall that in the model, the activity of a (topological) group of neurons is represented by images in terms of numerical values that express the probability of firing (action potentials) during a defined interval of time. The brain, of course, does not work like this. Neural activity is asynchronous, and all-or-none.

The firing probabilities, however, often can be related to a histogram analysis of neural activity [12.6, 20]. The representation captures the basic features of the neural activity, and can therefore be called relevant. In that sense, the auditory nerve images, as well as the onset images, are relevant representations of the signal at the peripheral level of the auditory system. It is less clear whether this concept of image is relevant at higher levels of the auditory system – where neurophysiological data are far less specific.

Next comes the question whether the tone completion image is relevant. J. Schouten (in 1940) was the first to demonstrate that the perception of the missing fundamental could not be an effect of the auditory periphery. Later on, it was demonstrated that the missing fundamental could be perceived when two different harmonics are presented dichotically [12.12]. This suggests that the pitch percept is located at neural centers which process information from both ears.

The first nucleus, which processes binaural information, is the medial superior olive. The temporal information processing of this center is known to be accurate [12.20] and fusion can happen in this nucleus [12.5]. Another probable candidate is the inferior colliculus (IC). The IC is known to display in its signal representation most of the temporal range that constitutes the residual pitch range. *Schreiner* and *Langner* [12.22] provide neurophysiological evidence suggesting that the IC plays a major role in the extraction or representation of temporal signal aspects in the residual pitch range.

Their method, based on the observation of the temporal characteristics of neurons (the temporal modulation transfer functions) in combination with the frequency tuning curve, provides a complete description of the response characteristics of neurons. Schreiner and Langner have found that temporal signal aspects such as envelope variations represent a major organizational principle of the auditory system in addition to its spectral organization. Neural elements tuned to different frequency ranges of envelope modulations are arranged in a topographic fashion within a coordinate system for the representation of certain temporal features of signals [Ref. 12.22, p. 357].

Schreiner and Langner's results show that the neurons of the IC are already organized topographically with regard to the signal features of the temporal domain (amplitude modulation). Furthermore, they found that the temporal resolution of neurons in the higher centers of the ascending auditory pathway is decreasing. They argue that since the upper frequency limit of temporal resolution for the majority of auditory cortical neurons is below 100 Hz, and since the majority of neurons are tuned to modulation frequencies well below 50 Hz, most signals with a periodicity or a residue pitch above 100 Hz (including most music and speech signals) appear to have no direct temporal correlates of their pitch aspects represented in the cortex.

According to Schreiner and Langner, there are indications that the IC plays a major role in the extraction or representation of temporal signal aspects in the residue pitch range in that it creates a spatial representation of periodicity information : "periodicity pitch is thereby probably encoded by a spatial arrangement of periodicity-tuned units" [Ref. 12.22, p. 358]. Their hypotheses are (1) periodicity (or residue) pitch in mammals, including humans, is analyzed in the time domain, (2) a spatial representation of periodicities is generated by a correlational analysis in the time domain, and (3) periodicity pitch is thus probably encoded by a spatial arrangement of periodicity-tuned units.

The results of Schreiner and Langner speak in favor of the place-time model and subsequent processing by a spatial model. It is important to notice that the model of tone semantics, in particular the connection between preprocessing based on VAM and the self-organization model SOM involves a transformation from the temporal to the spatial domain. The observations indeed support the view that correlational analysis in the time domain gives rise to a tonotopical representation of the residual tone patterns. The further processing at higher levels (self-organization) can then be assumed to be based on topographical features.

The neural correlation model proposed by Schreiner and Langner, however, does not exactly match the auto-correlation function. The model assumes two inputs: one input is synchronized to the temporal structure of the carrier signal (dependent on the best frequency of the auditory nerve fiber), the other is synchronized to the envelope of the signal (dependent on the best modulation frequency for the neuron). The output of the correlator has a high firing probability when both inputs occur simultaneously, but this coincidence condition depends furthermore on intrinsic oscillations occurring at periods of 0.4 ms [12.14, 15]. According to Langner and Schreiner, the correlation model accounts for the first and second effect of pitch shift.

Apart from the IC, however, there is also evidence that the perception of the residue tone is mediated by the auditory cortex. This evidence comes form studies in amnesia. Zatorre [12.30, 31] found that Heschl's gyri (the primary auditory cortex) and its surroundings in the right hemisphere play a crucial role in extracting the pitch of the missing fundamental. According to Zatorre, the observations, based on patients with right temporal lobectomy into which Heschl's gyri were excised, are compatible with the idea that the function of the central pitch processor is based on processes of pattern-matching.

Neurophysiological studies suggest that the perception of the residue pitch has a physiological basis in the temporal properties of neurons in the brain stem, but it is not excluded that cognition processes and pattern matching play a role as well. In Chap. 3 it has already been mentioned that voluntary aspects may play a role in the determination of the perceived pitch. It is indeed possible that the decision processes are located in the auditory cortex, while the mechanism that generates the residue pattern is provided by the brain stem nuclei. These observations are not in contradiction with the research of Schreiner and Langner.

A second question, and related to the first, is whether tone completion is learned or physiologically based. The place model and place-time model are based on different opinions on this subject. According to *Terhardt* [12.27], subharmonic patterns are learned early in live (even before birth), when the child is confronted with complex pitches in the surrounding world (e.g., the speech of the mother). During the learning process, the correlations between the spectral pitches of speech are recognized and stored (printed) in memory as subharmonic patterns. These patterns function as pattern completion devices in the way described earlier by the place model. A pattern completion model such as the learning matrix or a perceptron network could simulate such a completion device.

According to the place-time model, however, the subharmonic patterns are the result of the response characteristics of neurons. The correlation function is seen as a model of the probabilistic firing mechanism of neurons. An interesting consequence of the place-time model is that harmony could have been developed in a "pure tone" world – as van Noorden notices [12.29]. Recent research about inharmonic tunings support this view: the type of tone semantics that emerges from listening to Western music is one of a possible multitude of different tone semantics whose foundation is ultimately based on the acoustical properties of the sound and the temporal analysis by the auditory system. If sounds are used with stretched harmonics in stretched scales, under certain conditions, it is possible to produce aspects of tone semantics that are similar to the "classical" tone semantics (Cf. the Bohlen-Pierce scale [12.17, 18]).

12.2.2 Foundations of Schemata

The term *functional organization* means that neurons have a particular functionality or response characteristic for a given stimulus. For example, there is evidence that neuronal functions of different nuclei in the auditory brain are ordered according to a specific axis of frequency. These so-called *tonotopic maps* are logarithmically distributed and correspond to the place coding of frequency along the basilar membrane. Alternatively, this is called a *cochleotopic organization* of neurons.

The model of tone semantics relies on the notion of functional organization, albeit not necessarily a tonotopic organization – it may be called *chordotopic* organization. A distinction can furthermore be made between three types of organization: projection, self-organization, and association.

1. **Projection**. According to *Schreiner* and *Merzenich* [12.23], all nuclei of the ascending auditory pathway are cochleotopically organized. This means that the place coding is projected or reflected in the main neural centers in the brain. The temporal coding, on the other hand, is also

projected onto the different stages but the overall resolution decreases in successive stages. The responses of auditory nerve fibers show a high temporal resolution but a lower temporal resolution was found in the cochlear nucleus and the auditory cortex [12.22].

2. **Self-organization and Functional Maps**. Although tonotopic maps at the periphery of the auditory system reflect the ordering of the auditory nerves which have their receptors in the hair cells, it is not excluded that tonotopy at higher levels can be the result of self-organization. An example of such an interesting mechanism is given in *Kohonen* [12.13]. Still more interesting is the fact that there is evidence for a number of different types of maps, beyond the tonotopic representation. Among different species, auditory maps have been found that are very specific for e.g., amplitopic representation, odotpic ("echo delay") representation, Doppler-shift (frequency-frequency) representation, the representation of binaural data, space maps, amplitude modulation rate [12.22] and other. According to *Suga* [12.26], the size and topographic environment of these maps is an indicator of the importance of the parameters for the species. His work provides evidence for the existence of cortical maps for auditory imaging in the mustached bat (the Pteromotus parnellii). The mustached bat emits complex biosonar signals and listens to echoes for orienting itself and for hunting flying insects. These signals get localized somewhere on an internal map in the cortex. The map functions as a kind of resonance system in response to the environmental stimuli. Signals acquire meaning because they are relevant for the action of the organism in the environment.

   If such maps exist for dedicated functions and analysis of signals, it is probable that a similar structure (a schema) may exist for tone center analysis. In other words: if the organizational principles on which our study of self-organization is based, indeed has a neurophysiological foundation, then it may make sense to assume a place in the brain (of listeners to Western tonal music) where the functional organization of neurons has a response structure which is similar to the one obtained by the self-organization model. Similar observations can be made for timbre analysis.

   Listeners in other cultures are supposed to develop schemata whose functional organization may be different. Aspects of rhythm may indeed have a much larger impact on the tone schemata of non-Western cultures. As argued, the schema will depend on the interplay of acoustical features of the sounds of the instruments used, the auditory system and brain mechanisms (which are supposed to be invariant over cultures), and the distribution of the form bearing elements in space and time. The thesis that tone semantic functions have a specific location in the human brain of Western listeners is a working hypothesis. Up to now there is no direct evidence from brain research in support of such a functional organization.

3. **Auto-association.** The notion of self-organization, such as used in the previous paragraph, is restricted to *data-driven* learning processes. It was argued that tone semantics also involves *knowledge-driven* processes. In particular, a preliminary model of knowledge-driven tone semantics was worked out in the TCAD-model. With TCAD, an attraction dynamics of auto-association was simulated.

   The neurophysiological basis of auto-association is founded on the relations between neuronal functions and is often associated with an attraction dynamics. Without going into the details of this field it is instructive to make a further distinction between various mechanisms of auto-association. One is related to *interpretation*, while the other is related to *unification*. In TCAD, the concept of *interpretation* has been connected with a dynamics that is based on a reconsideration of the past. The model is highly speculative and not based on any known mechanism at the neuronal level.

   Unification deals with the problem of how information, analysed and represented by different functions in the brain, can lead to unified concepts. Since our model is basically restricted to only one such function, the problem of unification was not a central problem. It will become a central problem however, when other modalities, including rhythm and timbre perception are integrated into one single model. There is evidence that unification association is based on the correlated activity of neurons which belong to different functional entities [12.21]. Stated differently, unification would emerge from temporal signal correlations between neurons. The hypothesis that correlated activity between neurons is a vehicle by which attributes of a signal (processed in different specialized centers) are unified with other sensoric modalities may also be important for the future of music semantics.

To conclude: the neuro-sciences are an important source of inspiration for building models. Up to now, a high degree of E-relevance is found in the preprocessing stages and possibly also in the model of self-organization. The E-relevance of the TCAD model is less evident, partly because the model is still in a early stage of research, partly also because the neurophysiological foundations of cognition are still topical issues in research. Many authors agree, however, that progress in brain research can only be achieved when computer models of neuronal functions accompany empirically based investigations.

12.3 Modular Organization

According to the modularity thesis of *Fodor* [12.8], the cognitive system contains a number of distinguished subsystems with specific characteristics. The central processing system is specialized in association, symbolization and metaphorization, while the distinct subsystems (also called input systems or perception systems) are specialized in the detection of invariants and the formation of percepts and gestalts. Subsystems are domain-specific and can be associated with a neural system. A module has it own processing capacity and is not much influenced by other modules or by the central processing system. A module is mandatory (relying on the automatic processing of input) and the processing is very fast. It is informationally encapsulated and cognitively impenetrable. According to Fodor, the central processing system has access to the output of the module.

Whether the complex dynamics of music perception should be conceived of as a collaborating and mutual interaction between different such subsystems is not entirely clear. The above analysis, however, suggests that tone semantics may relate to one such specific subsystem. Perhaps also rhythm and timbre processing can be related to a specific subsystem. On the other hand, they should be conceived of as emerging aspects of the auditory system and, from this point of view, it is not entirely clear whether they form independent subsystems or whether they actually belong to one and the same subsystem (with internal differentiation).

Peretz and *Morais* [12.19] point out that researchers in the domain of neurophysiology have been reluctant to adopt the modularistic view. The research has been guided by the implicit idea that music perception involves the execution of general-purpose mechanisms. With new developments in schema theory this attitude may change.

12.4 Relevance for a Theory of Meaning

In connection with the problem of epistemological relevance, there are a number of important questions related to meaning formation and semantics. The computer model entails a semantic theory which by far extends the domain of pitch perception. To start with, it is interesting to relate this theory to two other theories of expressive meaning formation.

12.4.1 Expressive Meaning and Analogical Thinking

To exclude all misunderstandings: tone semantics theory is not about *expressive* meaning formation – such as formulated in *Broeckx* [12.1, 4].

According to Broeckx, the fundamental basis of (expressive) meaning resides in the fact that music is perceived as an acting organism to which sensory properties and events are assigned. Expressive meanings, according to this author, are the result of a highly abstract level of association, symbolization and metaphorization. These functions are related to what he calls *analogical* thinking. According to Broeckx, meanings are generated by *synesthetic* and *kinesthetic* processes which associate with auditory perceptions.

By means of *cenesthethic* processes, these associations can probe emotions and affects but this requires a dedicated listener. In that sense, meaning assignment is seen as the result of a cognitive process. The formation of the relationships between auditory perceptions and emotions are called *cognitive* because it is related to the way in which knowledge is acquired. In this process, the formal properties of the musical entities are the entities of the knowledge and their metaphorical transformation to the domain of emotion and affect is the subject of the knowledge process.

The theory of expressive meaning formation is not in contradiction with tone semantics theory because the latter is to be situated at a much lower perceptive/cognitive level. Tone semantics implies *automatic* and *direct* meaning formation, without much explicit conscious processing. Expressive meaning formation, on the other hand, is situated at a high cognitive level – which is *not* automatic and indirect.

12.4.2 Expressive Meaning and Virtual Self-movement

Another theory of expressive meaning formation has recently been formulated by *Todd* [12.28]. This approach is in many ways different from Broeckx's. Rather than associating the expressive meaning formation with an aesthetical experience, it is associated with physiological foundations for self-motion. Musical expression according to N. Todd is the result of a harmonic or rhythmic interpretation of music by the performer. The information provided by the performer is necessary to let listeners comprehend the complex and ambiguous organization of music. In this process, the performer first analyses the musical skeleton (the score) and codes the analysis in variables which he has at his disposal using changes of tempo, and accents in loudness, and pitch.

Musical expression is seen as the result of a mechanism by which these variables evoke the effect of self-movement. Research into the perception of tempo shows that expressive tempo-variation obeys laws which are similar to the laws of mechanics in which objects move under the forces of gravity. Given this analogy, expression is seen as the communication of self-movement. Expression creates the effect that the listener is virtually moved and this helps in solving ambiguities of the musical structure.

Contrary to Broeckx's theory, this theory is supported by a computer model. The mechanisms which are responsible for the perception of virtual self-movement are situated at a low perceptual level which, according to Todd is to be situated in the vestibular system. The effects are automatic and direct: the expressive meanings result from the way in which the perceptive system codes and processes the auditory information.

12.5 Music Semantics and Meaning Formation

Our thesis that tone semantics relies (a) on a perceptual schema which develops by self-organization, and controls (b) a dynamics of context-sensitive similarity relationships, is an example of meaning formation at a non-symbolic level. In what follows, tone semantics is the starting point for a general theory of musical meaning formation.

The starting observation for a theory of meaning is that a system may develop actions on the basis of an image (internal representation) of the world. Signals become meaningful because they are found to be relevant for the action of the organism with respect to a stimulus-context.

For example, a flute-signal during a football match has meaning for those who know the rules of the game. In the traffic context this same signal may warn the artless cyclist that he is committing a road offence. These are *learned* meanings with an arbitrary character. The signals are interpreted because they have been conditioned and associated. Direct or perceptive meanings, however, are not learned by convention. Biosonar-signals, for example, have meaning for bats, because they are relevant for the localization of preys. This type of meaning-assignment is not the result of arbitrary association, logical deduction, or inference, but of the way in which the animal responds to stimuli in the temporal and spatial acoustical and non-acoustical environment. This response is determined by neurobiological structures and *perceptual learning*.

A similar type of meaning formation occurs during the perception of tones. Tones have no meaning as individual elements but "get" or "resonate" meaning within a context. Tone functions are not learned by convention but emerge from the listening process. The meaning formation for tone perception is based on automatic processes. This is both based on learning and neurobiological conditions. Reference can be made to tone fusion theory, which states that tone fusion occurs from correlated neuronal activity of the temporal frequency encoding.

As the VAMSOM model shows, a schema can emerge from completion images, but the memory structure is highly cultural because the completion images depend on particular distributions of tones, and acoustical properties of the instruments. In this respect, the type of tone semantics that has been studied in this book is but one of many possible tone semantics.

In general, tone contexts are highly culturally determined and the semantic relationship of tones relative to each other can only be conceived of within the constraints that have been mentioned before. The emergence of meaning between tones (expressed in terms of the correlation between tone images) depends on acoustical properties of the musical instruments used, of the specific combination of tones, and all this is processed by the dispositions of the auditory system and the organizational principles of the brain.

The theory of meaning should therefore be conceived of within the global framework outlined. Central concepts of this theory are : direct perception, perceptual learning and context-dependency.

1. **Direct Perception**. The notion "direction perception" comes from *Gibson* [12.9, 10]. According to Gibson, invariants, necessary for the specification of all significant objects and events are directly extracted from the information stream that is presented to the sensorial system. The invariants correspond with the permanent properties of the environment. The function of the brain is to detect these invariants among the changing sensations. The extreme realism of Gibson's approach has recently been reconciled with the notion of "internal representation" [12.16, 24]. The neural network approach provides a reasonable basis for a Gibsonian perception theory.
2. **Perceptual Learning**. Direct perception is based on perceptual learning. This refers to the capacity in which a system may extract information from the environment. Since learning is involved, the mechanism can be fine tuned. What is learned are classes or invariants, i.e. things that have some constancy over a continuously evolving information stream. These classes can be interpreted as stable points. In tone semantics it is possible to show that such stable points may emerge from the self-organization of the context patterns.
3. **Context-Dependence**. The third and perhaps most important concept for a theory of meaning is context-dependency. The meaning of a perceived object changes depending on the context in which the objects appears. But the difficulty is that the context is determined by the object too. Our theory states that contexts are attracted by stable points that emerge on a long term scale. This tension forms the basis for the functionality and the attributed meaning.

The present theory combines the ecological theory of meaning [12.10] with the cognitive structuralist approach of *Shepard* [12.24, 25] and the theory of self-organization [12.11, 13].

12.6 Epistemological Principles

The theory of tone semantics covers only a small part of a more general theory of musical imagery. The case-studies in this book provide a non-symbolic paradigm for an operational account of semantics that will serve as the basis for future research. It is therefore important to have some insight into the epistemological principles of this approach.

A distinction is made between two paradigms for music research: the symbolic and the non-symbolic (also: sub-symbolic) paradigm. They are characterized by the following opposite principles:

12.6.1 Atomism vs. Continuity

Epistemological atomism is perhaps the most important property of the traditional account to music. The term *atomism* originally refers to the ancient theory that matter is not divisible ad infinitum, but consists of small particles that cannot be divided further, the so-called atoms. The epistemological variant of this theory states that the basic elements of musical knowledge consist of chunks, called knowledge atoms. These are believed to behave as semiotic entities with an independent and fixed status. Everything that can be observed, and in general any knowledge, can be reduced to (or deduced from) these elementary data (sensa). The knowledge atoms are therefore the most elementary data. They come into existence by a transducer mechanism (the ear), by which physical input is transformed into symbolic output (the atoms). Examples of this atomistic attitude are abundant in musicology. In particular we refer to those theories that take the note as the basic unit of knowledge.

In the non-symbolic account, atomism is replaced by continuity. Objects of musical knowledge are considered to be the result of a continuous processing of sensorial information. Musical information processing is based on the (automatic) detection of invariants and discriminating features in the stimuli. Hence, the representations are not propositional but have an analog and variable basis. The system is able to form an image of the external stimulus as a neural code, and any further processing is based on this. There are no predefined atoms or atomic sensations - except conglomerations that are activated. Any distinction is the result of a conceptualization.

12.6.2 Cartesian Dualism vs. Monism

The traditional approach gives a special status to the mental system. The mental system, which works on the atoms as an interpreter, has an autonomous status that is considered to be independent of a physical realization or implementation. Although the brain is the carrier of these processes, they can be described as if they exist in another (mental) world.

The non-symbolic account, however, sees the mental world as an epiphenomenon of the physical world. The mental system is identical to the neuronal system. The difference is but a consequence of the way in which this reality is conceptualized. For that reason this monism is also called psycho-physical parallelism.

12.6.3 Computational Formalism vs. Complex System Dynamics

The mental system has a formal and computational nature. Mental processes are formal processes and are to be expressed as rules that can (only) be applied onto the formal properties of the representation. Hence, mental processes have no access to the semantic properties of the representation. Semantics is something that comes from outside the representation.

To the contrary, the non-symbolic account sees perception and cognition as the result of dynamic interaction of complicated and connected mechanisms. All information processing activity is based on the global dynamics of the interaction between simple local processing units (the neurons). High-level concepts emerge from this on the basis of self-organization. Semantics should be conceived of in terms of complex dynamic systems. Percepts and concepts are conceived of as attractor points, that is, stable points of the system state, the meaning of which should be understood in terms of the interaction with the environment. Fluctuations in the environment or even in the system itself may cause a transition of the stable points.

12.6.4 Representational Realism vs. Naturalism

Realism refers to the view that there exists a real world independent of our thought. According to the symbolic account, the mental representations must be connected somewhere to the real world because the representations "represent". The symbols that designate musical objects in the world represent because some causal connection between the objects and the representation (cf. the transducer) should exist. However, we do not need to establish such a connection to be able to work with it ("Methodological solipsism").

The non-symbolic account takes the brain as the basis for the simulation of intelligent behavior and states that there must be some causal connection between the acoustical environment and the representation in the brain. Naturalism states that there are no other entities, objects or events than those given by nature and those that can be accessed by scientific methods. According to this view, the connection between a mental representation and the real world can be provided by a model of the sensorial and low-level perceptual mechanisms of the auditory system.

12.6.5 Methodological Solipsism vs. Methodological Ecologism

The symbolist holds the opinion that although an interaction between the world and the organism is supposed, it is not necessary to simulate this interaction in the praxis of everyday cognitive research [12.7]. The character of the musical processes is just formal and therefore independent of the causal factors by which the symbols, as representations of the environment, come into existence. The semantic interpretation of these forms is done by a user of the formal system. The programmer is like Descartes' God, who implants the concepts and their relations into the memory of the knowledge systems and interpret the symbols he sees. God is necessary because, without him, this implanted knowledge has no meaning at all and nobody would be able to know how this system came into existence. Hence, the claim that the world is produced by the mind of the individual subject is accepted only for methodological reasons. In essence, the symbolist is a realist.

According to the non-symbolic account, the interaction with the world should be implemented because this is the starting point of all knowledge at higher levels. Rather than being solipsist for methodological reasons, the non-symbolist is ecologist for methodological reasons. The model of a cognitive process is therefore also a model of the environment in which this cognitive process operates.

12.6.6 Cognitivism vs. Materialism

The symbolic approach states that cognition has a status independent of the physical carrier of the system. The basic processes can be expressed in the language of formal logics. The symbol is the carrier of any kind of musical information, but it has an arbitrary character with respect to any designated object. The semantics of the symbol is mediated by thc intentional attitude of the system, the interpreting system that is present in any human. This cannot be simulated with digital computers, but it is not necessary. We do not need it for a theory of musical information processing.

According to the non-symbolic approach, the information processing system is not a formal system and does not work as an autonomous formal system on implanted concepts. All musical knowledge is developed by learning, which means that any knowledge is the result of three factors: the environment (comprising the distribution of the information), the physical properties (of the information and the brain), and the dynamic properties (of information processing in the brain and of the action of the system in the environment). These properties have to be investigated by scientific investigation into the neural system and the theoretical study of dynamic systems. The non-symbolic account does not exclude the possibility that more global concepts can arise. Yet, these are the result of processes of self-organization. Self-organization provides the basis for a causal theory of perception and cognition. Hence, the basic foundation is materialistic.

Research in music recognition and listening may be called prototypical for the relevance of this epistemological analysis. Indeed, what constitutes a psychologically relevant semiotic unit is not unambiguously predefined in this field. Unlike speech-recognition research, we can not even dispose of templates that would provide musical equivalents of phonemes or words. Auditory concepts emerge in our imagination as a part of and during the complex dynamic processes. It makes little sense to found a theory of cognition on abstract implanted concepts.

12.7 Conclusion

The schema theory proposed in this monograph entails a psycho-morphological theory of semantics. It is based on the distinction between two types of dynamics. The first type is on a long term scale. As mentioned before, it results in a more or less stable perception structure. The second type is short term. Input patterns that stand for percepts (such as tones, and chords) are correlated with integrated patterns that establish a musical context in the stable memory structure. The meaning of a tone or chord then emerges from the tension between processes that occur at different time scales. Tone meaning is in essence defined as the tension created between a particular percept (of a tone or chord) with respect to a spatial-temporally bounded context in a stable perception structure. The approach shows ways of quantifying this tension and this has been illustrated in tone center recognition of music. In that sense, the theory of tone semantics is much more than a theory of tone center perception because it includes the notion of contextual dynamics and perceptual learning. Extension to music semantics and semantics of perception in general is straightforward.

13. Cognitive Foundations of Systematic Musicology

This chapter discusses the historical-scientific context of systematic musicology and it relates the paradigm which has been developed in the previous chapters with a psycho-morphological approach of music. The chapter closes with a specification of the interdisciplinary and cognitive foundations of systematic musicology.

13.1 Cognitive Musicology, AI and Music, and Systematic Musicology

In the last decade, Cognitive Musicology (CM) and Artificial Intelligence and Music (AIM) have come into the focus of advanced music research [13.3, 5, 15, 18–20, 22, 23, 33, 41]. Several authors have suggested that the world-wide interest in musical cognition is the outcome of developments in the late 60s and 70s [13.17, 40]. In particular, these authors mention the influence of cognitive science, structuralism, and developments in computer technology. Yet, music cognition has since long been a central research topic of the European tradition in systematic musicology. From this viewpoint, the originality of CM and AIM should perhaps not be exaggerated as the early systematic musicology is in many respects a cognitive musicology "avant-la-lettre". The revival of systematic musicology in Europe can be related to the new successful development of CM and AIM.[1]

[1] The *First International Symposium on Systematic Musicology* was organized in Bratislava (1993) and a second edition (with a focus on cognitive foundations) followed in Hamburg (1994). The *First International Conference on Cognitive Musicology* (Jyväskylä 1993), was the first conference in which the term "Cognitive Musicology" was used. Yet, since 1987, a number of workshops on CM and AIM have been organized in Europe and the USA. Some of these workshops took place within the context of large AI-conferences such as ECAI (Stockholm 1990, Vienna 1992), or AAAI (Minneapolis/St.Paul 1988), IJCAI (Detroit 1989), ... Other initiatives were taken on an independent basis (Ghent 1987, St.Augustin 1988, Marseille 1991, ...). Italy has a national conference on Music and Informatics. There are also the conferences related to computer music (ICMC) and psychology (Paris 1988, Liège 1994). Although this list is not exhaustive, it suggets that CM and AIM have been important for the developments in systematic musicology.

13.2 Historical-Scientific Background

The term "Systematische Musikwissenschaft" goes back to Adler, who made a distinction between historical and systematic musicology [13.1]. The historical part consists of musical paleography (notation), fundamental questions in history (e.g., about the grouping of musical forms), the historical study of style, and organology (history of music instruments), while the systematic part deals with research and foundations of the art of tones ("Tonkunst") (in harmony, rhythm and melody), esthetics, pedagogy and didactics and ethnomusicology. Disciplines, such as acoustics, mathematics, physiology, psychology of tones, and logics (musical thinking) are considered help sciences.

Adler's definition has a broad scope, and an interpretation exists which is more oriented towards psychology. *Wellek* [13.42], for example, pointed to physical and physiological acoustics, psychological acoustics, and music psychology as the basic disciplines of systematic musicology. This viewpoint entails that musical esthetics, philosophy, sociology and pedagogy are independent disciplines which have a place within the broad interpretation of the concept of "Systematik". In the narrow interpretation, research is focused on the foundations of "Tonkunst" (tone art) – a subject related to the problem of tone semantics and composition theory.

The discipline started with the major contribution of Helmholtz and evolved at the end of the 19th century towards an independent research program in music cognition. The central thesis was that [Ref. 13.42, p. 5]:

> music is neither a physiological or physical fact, but a psychical and mental one: phenomenon and structure – the first as experienced and listened, the second as work, created by people.

The program envisioned the analysis of sensorial, perceptive and cognitive processes with a focus on the psychological aspects.

Authors such as Riemann and Stumpf saw this program realized in terms of a *tone epistemology.* It included the study of tone imagery, representation and consciousness. According to *Stumpf* [13.37, 38], musical information processing results from the logical activities of the "Anschauung" (apperception). Tone consciousness emerges from (cognitive) fusion processes, whose foundation is physiological.

Other authors, on the other hand, argued that the tone epistemology developed by Stumpf was too narrow. Musical semantics was not merely limited to tones, but included forms (Gestalts) and structures based on rhythm and timbre patterns. The term "psychology of hearing" ("Gehörpsychologie") was suggested instead of "tone psychology" [13.42].

The foundations of this musical epistemology were influenced by the main philosophical currents at the time: *idealism* and *phenomenology.* In the works of Riemann and Kurth, for example, the foundation is idealistic and vitalistic. Listening is considered as a highly developed skill of *logical functions* of the human mind [Ref. 13.8, p. 47] or (in Kurth's vitalistic approach) as a subject of *energetic forces.* Stumpf and Wellek base their theories on phenomenological and Gestalt theoretical accounts.

Looking back, it appears that the interest and research into musical cognition is certainly not a recent phenomenon: already in the 19th century it *was* the central research subject of the "Systematische Musikwissenschaft" [13.13, 34, 32].

13.3 New Developments in the 1960s

In the early 60s, however, a paradigm shift can be observed. Wellek is often considered as the last representative of the research program in systematic musicology. Since that time, the cognitive musicology "avant-la-lettre" developed into a programmatic enterprise or evolved towards a musical hermeneutics in which the disciplines of the periphery such as esthetics, philosophy, and, especially, sociology and pedagogy, shifted towards the center. The broad cultural and sociological context came into the foreground and the original connection with psychoacoustics got lost – although the field of "Gehörpsychologie" (psychoacoustics) was developing fast [13.43].

One of the aspects that probably explains this development has to do with the internal history of psychoacoustics. With the development of electroacoustics, this discipline became an independent and highly specialized research area, whose foundation was in disagreement with the Gestalt psychological and phenomenological basis of music cognition. The empiricist approach assumed that knowledge is deduced from the sense data provided by the auditory channels, whereas the Gestalt theory assumed that perception is based on autonomous forces (including learning processes) residing in the brain. As far as methodology was concerned, psychoacoustics focused on a stimulus-response paradigm (registration of response on artificial stimuli) whereas musicology didn't have an instrument to integrate the strong empiricist approach within the global approach of the Gestalt theory. Nourished by the influential writings of Riemann, who cultivated a certain aversion towards empirism, musicology remained a descriptive (qualitative) and highly programmatic [13.33] science that failed to integrate the new developments in auditory perception.

Another compelling factor in the paradigm shift was the evolution of musical composition itself. New developments in electro-acoustic music since 1950 have forced the researchers to adapt and device new methods. The tone was no longer considered the atom of musical composition: noises, timbres, clusters, concrete sounds, and even ... *silence*, became new building blocks. The latter was conceived of as an open window to the world of the extra-musical, the opening of form and content. Art and music became "open" and together with the tone, also structure disappeared. A piece of music became not only logically indifferent, but also chronologically indifferent. Sabbe [Ref. 13.31, p. 177] says:

their being no organizational hierarchy whatsoever among the different structures and therefore no logical reason for any kind of precedence, there is no logical foundation for any kind of positional precedence over time for any one of the structures, either. So, the principle of the circularity of musical thinking, the type of causality of which is expressible only in terms of cybernetics, and its reconciliation of irreversibility and retroactivity – all of which were present "in nuce" in late Beethoven – ended up in open form aleatorics as introduced in European music in the second half of the fifties.

The researcher adapts his methods in an attempt to understand (or to let understand) these developments. He sees that the traditional approach is inadequate and searches for new means to study these phenomena [Ref. 13.9, p. 10].

Summarized, there are internal as well as external factors that have contributed to a shift from a psychological based systematic musicology towards a sociological based hermeneutic musicology. This orientation takes place in the early 60s, when the concept of "Tonkunst" is replaced by a broader humanistic and sociological approach.

13.4 A Discipline of Musical Imagery

The situation has now changed. First of all, in psychoacoustics, the boundaries between sensory, perceptual, and cognitive processes have become diffuse. Psychoacoustics is no longer related to purely sensory phenomena but accounts for auditory phenomena in general, including the so-called Gestalt psychological aspects of perception such as contextual meaning formation, auditory fusion and segregation. Of particular relevance at the time (in the 1970s) were the pitch perception models. Terhardt's virtual pitch theory (Sect. 5.4) assumes that the perception of pitch results from an analytic analysis in the auditory periphery and a global analysis of central origin. The latter is conceived of in terms of a pattern-recognition system based on learned templates of subharmonics. Its foundation is assumed to be cognitive rather than sensorial or perceptual. And, although the cognitive status of the subharmonic templates can be questioned, the general idea that perception often involves learned categories remains an important one. It has opened ways to connect psychoacoustics with Gestalt theory, as studies in sound segregation clearly show [13.4].

The development of cognitive structuralism in the USA [13.36] has announced the start for new directions in music psychology, among which the exploration of auditory illusions, tone perception and timbre perception are most prominent [13.25]. The study of auditory illusions has contributed a lot to establishing a new paradigm of music research.

At the time of the decline of the traditional European Gestalt psychologically and phenomenologically-based systematic musicology, there were signs that a new approach was in the making – one in which the old phenomenal concepts of pitch, timbre, and rhythm were seen as emerging outcomes of complex processes of the human auditory system.

The present study therefore entails both a break and continuation with the tradition of systematic musicology.

- A break because the qualitative and descriptive character of the phenomenological and Gestalt psychologically based paradigm is now replaced by a quantitative (computational), empirical and model based (computer based) approach.
- A continuation because the problems of Gestalt perception and context-sensitive meaning formation are still considered core problems in cognitive musicology and cognitive psychology.
- There is also a link with the hermeneutic approach in that music semantics is believed to evolve within a typical historical context.

Given this background, Shepard's study of auditory illusion [13.35] gets a special status as the paper introduces a totally new perspective on music cognition at the time where the systematic musicology comes to an end – it is a perspective in which the computer plays a role:

1. as an instrument for the synthesis of sounds (this gives the researcher an absolute control over all parameters of music),
2. as an instrument for the processing of data (e.g., multi-dimensional scaling analysis, statistical analysis, ...),
3. as an instrument for the simulation of complex dynamic processes of the human brain (auditory models as well as neural models).

Shepard's paper announced the decline of the two-component theory and provided new evidence in favour of the view that phenomenal percepts, such as pitch, timbre and rhythm are emerging outcomes of complex processes of the human auditory system. The so-called illusions and paradoxes of music perception point to distinct mechanisms that often work in competition with each other (Cf. the perception of low pitch and high pitch). With the technique of multidimensional scaling, new insights were acquainted about schema based perception. In that sense, Shepard's article announced the start for new developments in musicology and it has opened the door for cognitive structuralism and a new theory of musical imagery.

Nowadays, music production again tends to focus on integration and construction [13.2]. With the advance of powerful computers, there is more than ever a need to base composition on auditory constraints. Composers acknowledge the tonal disposition of the auditory system and try to understand the underlying neurophysiological principles in order to explore new ways in post-tonal thinking. The explorations of the diffuse transitions between harmony and timbre, rhythm and tone, as well as stretched scales and related harmonies, has activated the interest in the micro-level representations of music [13.24].

As the subtilities of human perception are becoming more important in composition [13.10, 26, 28, 39] psychoacoustics gains new interest. Music may well be *logically* indifferent, and *chronologically* indifferent but not *psychologically* indifferent. In addition, as Risset notices [Ref. 13.27, p. 12]:

> The limitations are no longer technical, stemming from hardware problems, but intellectual, related to the software, the data bases, and the know-how.

In short, there are signs that the interest in cognitive musicology is related to shifts in music production: revival of tonality and the exploration of the transition borders of harmony and timbre, and rhythm and tones. The revival of tonality, however, should be interpreted as a renewed interest in how tone fusion manifests itself in timbre, harmony and polyphony. Most of the contemporary composers acknowledge the tonal disposition of the auditory system, although they don't want to be pinned down to a method of tonal composition [13.16]. So, it is not really a revival of the tonal system, but of the interest in tone center theory and timbre. Sabbe calls it "Obertonalität" [Ref. 13.30, p. 31]. It is determined by the sounds, the constraints of the auditory system, and the way the sounds are combined (diachronic as well as synchronic).

To conclude: the developments in music production and musicology itself have led to a new orientation of musicology. The study of musical imagery "as an internal activation of schemata, embedded into the general mimic activity of the organism" [13.21] became a central issue. The research subject is related to an important tradition in musicology but imports with it a new language of auditory/neural theory and powerful technologies for simulation.

Indeed, there is more than ever a new hope to give music theory and musical practice a foundation in terms of a neurophysiologically-based Gestalt theory. This is not a reductionist paradigm because it aims to explain how globally ordered structures and objects at the macro-level emerge from the interaction of elementary processes on a micro-level, and how in turn the behavior of the elements at the micro-level are influenced by the global macro-behavior. The current paradigm is based on a combination of self-organization theory (complex dynamic system theory) and physiological acoustics.

13.5 A Psycho-morphological Account of Musical Imagery

Psycho-morphology deals with the question of how form and structure develop from perceptive and cognitive processes, and how it influences these processes by feedback. It is a modern version of a psychoacoustically-based Gestalt theory whose aim is to explain how global order at the macro-level emerges from the interaction of elementary processes on a micro-level and how in turn the behavior of the elements at the micro-level are influenced by the global macro-behavior. Its dynamics is interpreted as self-organization theory.

Of particular relevance is the modular and accumulative method which is made possible by the computer. Indeed, models of the auditory periphery can be considered building blocks that one can use as foundations for music models to implement cognitive principles. Researchers in music do not have to model the auditory system from scratch but they can rely on work done by colleagues in psychoacoustics or engineering. In a similar way, results obtained in music perception can perhaps be combined and integrated into a larger system: an intelligent music workstation [13.6, 14]. The interdisciplinarity is not just in the banner but it is at the core of any project in this field.

The development of a psycho-morphological theory of music may have a large impact on other, more traditional domains of music research, such as music history. Tendencies and styles can indeed be seen as an emergent outcome of the activities of individuals in a particular socio-economic and cultural context. Once these individuals observe a global tendency, it may influence their activity to reinforce the tendency so that the general dynamics converges towards an artistic output with typical characteristics, called style, or system.

For example, it would be interesting to know how the use of the voice, certain instruments, tunings and playing techniques have contributed to the development the tonal system in Western Europe. The particular type of tone semantics of the West is indeed the outcome of a historical dynamics and it is important to study the development of schemata for tone perception within this broader historical context. A first attempt in this direction has been made by *Eberlein* [13.11, 12]. From this point of view, the extension from psycho-morphology to histo-morphology is a matter of time scales. Little research has been done to model historical dynamics within the psychoacoustic-based framework but the interest is growing.

On the other hand, one should also be aware of the limits of the present paradigm and perhaps try to anticipate some misunderstandings. A psycho-morphological theory of music is promising in that it takes full account of music as an acoustical phenomenon embedded in the constraints of sensorial, perceptive and cognitive processes. But a fundamental question is to what extent it is possible to relate this approach with the social-cultural environment outside music. Ethnomusicologists often point to the fact that extra-musical factors, which belong to the broader socio-cultural and economical context, play a major role in musical cognition. The present approach provides no straightforward answer as to how to bridge the gap between musical and extra-musical factors of musical cognition. It assumes that the intra-musical foundations of pitch, tone center, rhythm, timbre and sound-segregation are sufficiently constrained and defined.

13.6 Interdisciplinary Foundations

In an attempt to describe the current tendencies in music research, it is useful to distinguish between cognitive musicology (CM), and artificial intelligence and music (AIM). Both approaches relate to each other as two sides of the same coin but CM is more oriented towards problems in psychology and epistemology, while AIM is more directed towards practical applications. CM comprises the study of how musical knowledge is processed in mind and brain, while AIM is more directed towards the development of intelligent systems for different practical usages in the field of composition and analysis.

Yet, the difference is not always clear. For the simulation of aspects of mental and neural processing, for example, one often relies on concepts and techniques that have been developed in engineering approaches. On the other hand, there is evidence to assume that future "intelligent" music workstations will have to account for the psychological and epistemical constraints of human musical faculties: for example, like systems that extract information about tone, tone center, and beat.

With this book we have tried to show that the core problem of CM is related to psycho-morphology: the way in which forms (images) are manifesting and constituting themselves, emerge and organize. The core problem of AIM is more related towards applied morphology: the way in which forms can be used, for example in an interactive music workstation or multi-media computer [13.7, 29].

To summarize, the cognitive foundation of systematic musicology relies on the following disciplines: psychology, information science, musicology, philosophy, and brain sciences.

- **Psychology**. The aim is to quantify the relationship between man and music from different points of view: from the psychophysics of sound up to cognitive psychology. This discipline provides basic empirical facts about sensorial, perceptive and cognitive information processing in different areas such as listening, composing and performing.
- **Information Science**. This field includes all aspects that are of any help in modelling the relationship between man and music. It includes, mathematics, computer engineering, artificial intelligence techniques, logics – in short, all kinds of concepts, techniques and strategies for understanding and simulating information processing and dynamics involved.
- **Musicology**. This discipline studies the relationship between man and music from a phenomenological (intuitive) point of view, based on practical experience and historical analysis. The theories (from music theory up to esthetics and music history) provide an important source (sometimes

even formalized) of the most diverse aspects of musical communication and cognition.
- **Philosophy**. The modelling of the relationship between man and music poses important questions about old problems in philosophy, such as: how does knowledge comes into existence, what is perception, what is meaning. Philosophy provides a conceptual framework for a scientific analysis of the paradigm and the epistemological relevance of cognitive models.
- **Brain Science**. Brain science covers experimental as well as theoretical work. Physiology, neurobiology, and the new field of theoretical neuroscience provides data and insights into the mechanisms that underlay all musical activity.

13.7 General Conclusion

Music and schema theory have been studied in this book from the viewpoint of auditory models and principles of self-organization. Our aim has been to show that such an approach may provide a solid foundation for systematic musicology.

The focus has been on tone center perception and aspects of rhythm perception. This, of course, covers but a limited range of human musical activity. Less attention has been paid to the semantics of timbre, a unified model of pitch, rhythm and timbre, or aspects of segregation and integration. Furthermore, all those aspects that deal with social and cultural contexts, including the application of our model to non-Western music, have not been considered. In that sense, the present model is limited to "pure sound" cognition of Western music: those aspects of music cognition that are not contained in the sound itself have not been taken into account.

Further research will be needed to refine the model, to account for fully integrated modes of perception, to apply the model to non-Western music, and to integrate the extra-musical contexts in music cognition. This is not the task of one individual but, perhaps, of a community of systematic musicologists.

A. Orchestra Score in CSOUND

The following is a useful CSOUND-setup for generating Shepard-tones. CSOUND runs from two basic files: an *orchestra* file and a *score* file. The orchestra file contains an instrument (or set of instruments) that tells how to synthesize sound; the score file contains control parameters of which note should be played and at what time. In this example, the instrument produces Shepard-tones. Comments begin with a semicolon (;).

A.1 The Orchestra File

```
sr = 22050
; the sampling rate used by most DA-converters
; (Remind that the signal should afterwards be converted to 20000 to be used with
VAM)
kr = 551.25 ; control rate is 551.25 Hz
ksmps = 40 ; number of samples in a control period (sr/kr)
nchnls = 1 ; number of channels of audio output


instr 1 ; instrument number

;ktempvarN is a variable which is updated at the control rate (kr)
;The updating is guided by tablei which takes values from a table
;containing a bell-shape (see the function f4 in the score file).
;The place at which to take the value depends on pitch and this is
;calculated by means of a scaled logarithmic function.
ktempvar1 tablei ((log(cpspch(p5)/16)/0.3)-3.5)*32, 4, 0, 0, 0;
ktempvar2 tablei ((log(cpspch(p5)/8)/0.3)-3.5)*32, 4, 0, 0, 0;
ktempvar3 tablei ((log(cpspch(p5)/4)/0.3)-3.5)*32, 4, 0, 0, 0;
ktempvar4 tablei ((log(cpspch(p5)/2)/0.3)-3.5)*32, 4, 0, 0, 0;
ktempvar5 tablei ((log(cpspch(p5))/0.3)-3.5)*32, 4, 0, 0, 0;
ktempvar6 tablei ((log(cpspch(p5)*2)/0.3)-3.5)*32, 4, 0, 0, 0;
ktempvar7 tablei ((log(cpspch(p5)*4)/0.3)-3.5)*32, 4, 0, 0, 0;
ktempvar8 tablei ((log(cpspch(p5)*8)/0.3)-3.5)*32, 4, 0, 0, 0;
ktempvar9 tablei ((log(cpspch(p5)*16)/0.3)-3.5)*32, 4, 0, 0, 0;
```

```
;The variable actrl is used to change the envelope at the audio sampling rate.
;The signal generator expseg traces an exponential onset and offset of each 0.03 s
actrl expseg 0.01/p4,0.03, 0.5/p4, p3-0.06, 0.5/p4,0.03,0.05/p4;

;The variables asigN contain the actual signals.
;The amplitude of the signal depends on the onset-offset times (actrlN)
;as well as the spectral filter (ktempvarN)
asig1 oscil actrl*(exp(ktempvar1)*2000),cpspch(p5)/16,1;
asig2 oscil actrl*(exp(ktempvar2)*2000),cpspch(p5)/8,1;
asig3 oscil actrl*(exp(ktempvar3)*2000),cpspch(p5)/4,1;
asig4 oscil actrl*(exp(ktempvar4)*2000),cpspch(p5)/2,1;
asig5 oscil actrl*(exp(ktempvar5)*2000),cpspch(p5),1;
asig6 oscil actrl*(exp(ktempvar6)*2000),cpspch(p5)*2,1
asig7 oscil actrl*(exp(ktempvar7)*2000),cpspch(p5)*4,1;
asig8 oscil actrl*(exp(ktempvar8)*2000),cpspch(p5)*8,1;
asig9 oscil actrl*(exp(ktempvar9)*2000),cpspch(p5)*16,1;

;out sends the signal to the output
out asig1;
out asig2;
out asig3;
out asig4;
out asig5;
out asig6;
out asig7;
out asig8;
out asig9;

endin
```

A.2 The Score File

```
t 0 60 ;the tempo at time 0 is 60

f1 0 256 10 1
;f1 invokes a sinewave function table from which
;the numbers of the waveforms are read.
;It starts at time 0 (p2) and uses a table of 256 points (p3).
;The type of the invoked generator (p4) is 10 (GEN10) and only
;one single sinewave is used (p5)

f4 0 256 8 0 128 1 128 0
;f4 invokes a curve which is used for attenuating the
;octave-components of the Shepard-tone.
```

```
;It starts at time 0 (p2) and uses a table of 256 points (p3).
;The generator is of type 8 (GEN08) (p4).
;The fields p5, p7 and p9 specify the ordinate values of
;the function, while p6 and p8 specify the length of each
;segment. GEN08 constructs a stored table from segments of
;cubic polynomial functions, and the common slope is that
;of a parabola. This function defines the bell-shape for the
;spectral filtering of the octave components.

;The next lines specify when and which Shepard-pitches are to be played.
;The example specifies the Shepard-chord sequence CM—FM—GM—CM
;The first field (p1) contains the instrument number (i1).
;The second field (p2) contains the begin time of the note (e.g., 0.700000).
;The third field (p3) contains the duration of the note (0.500000).
;The fourth field (p4) contains the number of notes in the chord (e.g., 4).
;(the latter is used to obtain the same loudness for chords that have
;three of four notes)
;The fifth field (p5) contains the pitch specified as octave and pitch-class,
;thus 8.00 means the pitch DO on the 8th octave.
;The octave-components are generated by the instrument
i1 0.000000 0.500000 3 8.00
i1 0.000000 0.500000 3 8.04
i1 0.000000 0.500000 3 8.07
i1 0.700000 0.500000 3 8.00
i1 0.700000 0.500000 3 8.05
i1 0.700000 0.500000 3 8.09
i1 1.400000 0.500000 4 8.02
i1 1.400000 0.500000 4 8.05
i1 1.400000 0.500000 4 8.07
i1 1.400000 0.500000 4 8.11
i1 2.100000 0.500000 3 8.00
i1 2.100000 0.500000 3 8.04
i1 2.100000 0.500000 3 8.07

e ; end of score
```

B. Physiological Foundations of the Auditory Periphery

This appendix aims to give a short introduction to the physiological foundations of the auditory periphery. The ear is a sensor and a transducer of vibrations. Small changes in the atmospheric pressure are captured by the eardrum—the beginning of a long chain of complex events which ultimately leads to the perception of sound. The neural connections in the auditory system are relatively simple (compared to the eye), but the transduction from soundwaves to neural impulses is performed by a hydro-dynamic system with extremely complex properties.

B.1 The Ear

The ear consists of three parts: the outer ear, the middle ear, and the inner ear.

B.1.1 The Outer Ear

The outer ear consists of a shell and an ear channel, which is closed at one end by the ear drum. This part of the ear is more or less like a half-open tube in which resonances can occur at multiples of a quarter length of the waveform $\lambda/4$ (where λ is one period of the waveform). The outer ear acts as a filter: taking into account the properties of the tube resonances are found at about 2.6 kHz, with a gain in sensitivity of about 15 to 20 dB. The outer ear plays an important role in determining the direction of the sound source. It also produces a spectral modulation of the incoming sound.

B.1.2 The Middle Ear

The middle ear is about 2 cm^3 in size and contains three bones: the hammer, anvil, and stirrup which function as a lever by which the vibrations of the ear drum are transmitted to the oval window (the entrance of the inner ear).

Vibrations in one medium (the air) are transformed into vibration in another medium (liquid). Due to the power gain of the lever system the middle ear, too, acts like a filter: the maximum increase in sensitivity is about 28 dB, with a bandpass characteristic at about 1000 Hz.

B.1.3 The Inner Ear

The inner ear contains a complex of channels, called the labyrinth. The auditory part of the inner ear is the *cochlea*. There is also the *vestibular* part, which is mainly used for movement and sense of equilibrium. The cochlea is filled with liquid and has the form of a tube rolled up in the form of a spiral. The length of the tube is about 3.5 cm, with a diameter of about 4 mm^2 at the oval window (the base) and about 1 mm^2 at the end (the apex). Apart from the oval window, there is a second window (the round window) which closes the cochlear bone so that pressure in the tube (caused by the vibrations of the oval window) can be released at the other end by means of the round window.

Inside the cochlea, there is a sophisticated hydro-mechanical system which transforms the changes in pressure into electro-chemical impulses. Two basic structures should be distinguished: the *cochlear partition*, and the *organ of Corti*.

- **The Cochlear Partition**. The cochlear partition is the part of the cochlea between the *scala vestibuli* and the *scala tympani*. Due to the mechanical vibrations caused by the oval window, changes in pressure are generated in the cochlea and traveling waves are generated in the cochlear partition. Dependent on the temporal pattern of the movement, the waves generate a maximum amplitude of the partition at defined places: high frequency (=rapid movement) at the base, low frequencies at the apex. As such, a temporal pattern is transformed into a spatial-temporal pattern. At the places of maximum amplitude of the cochlear partition, sensors are excited which transduce the temporal pattern into electro-chemical impulses. An important characteristic of this transduction is that neurons tend to synchronize with the temporal excitation. (Both aspects of the encoding—spatial and temporal—are described below.)
- **The Organ of Corti**. The sensory structure on the cochlear partition, which transforms the mechanical energy into electro-chemical energy, is called the organ of Corti. On the top part of this organ are *hair cells* (inner and outer hair cells). There are about 3400 inner hair cells and about 15000 hair cells in one cochlea. The hair cells are terminators of neurons, whose cell body (*soma*) is located in the *spiral ganglion* (a collection of neuron cells which is located along the spiral structure of the cochlea). There are about 30000 such afferent neurons for one ear: they send auditory information to the central system. There are about 1800 efferent neurons by which the central system can influence the cochlea. On top of the hair cells are *stereocilia*, which transform movement into electro-chemical impulses.

B.2 The Neuron

B.2.1 Architecture

Neurons have a cell body and are connected to other neurons by means of an *axon* and *dendrites.* The basic structure of the neuron is such that it receives excitation from other cells or receptors by means of the dendrites, while the axon is sending the processed information to other cells. Neurons are information units of the brain and the connection and transformation of information between neurons or between receptors and neurons is realized by means of *synapses.*

A neuron can charge or decharge, provoking an impulse (*spike*) or *action potential.* A spike is an all-or-nothing event and the stimulus must be strong enough in order to pass the threshold potential. Once the neuron has fired, some time is needed to recharge. This delay period is called the *refractory delay.* The *absolute* refractory delay (during which the neuron cannot recharge) is about 1 ms, so that the maximal firing is about 1000 spikes/s. There is also a relative delay, during which the threshold is raised: the stimulus must then be stronger than normal in order to decharge the cell.

B.2.2 Analysis of Neuronal Activity

The analysis of recorded neuronal activity is normally done by means of a histogram. There are different types of histograms but all assume that time is made discrete and that multiple records must be made in order to obtain a reliable statistics. In the *interval* histogram, one measures the time between the subsequent spikes and holds a record of how many times each interval has occurred. In the *post stimulus time* histogram, one measures the time between the stimulus and the subsequent spikes. When the frame of the analysis is limited to only one period of the stimulus, then this is called the *period* histogram.

B.3 Coding

The hair cells are receptors of cells whose soma is located in the *spiral ganglion.* The axons of these neurons form the *auditory nerve* which connects to the *cochlear* nucleus—the first relay station. Due to the membrane properties of the spiral ganglion cell, the graded activity of the inner hair cells generate all-or-non activity in the auditory nerve fibers that innervate the cells. The auditory nerve fibers are tick enough to record and their response structure is therefore well known.

B.3.1 Spatial Coding

Depending on the temporal pattern of the signal, a traveling wave pattern is generated in the cochlear partition which produces an excitation at a corresponding place along the basilar membrane: low frequencies are encoded near the apex of the cochlea, while high frequencies are encoded near the stapes. The excitation patterns are such that tones may mask one another, or may provoke pitch shifts. Measurements show an almost perfect correspondence between the excitation of the basilar membrane, the hair cell responses, and the responses of auditory neurons.

B.3.2 Temporal Coding

The movement of the cochlear partition is transduced into neuronal activity. Graded activity of hair cells is transformed into the all-or-non activity of the action potential. The neuronal firing is in synchronization (phase-locking) with the positive phase of the stimulus. Only the positive phase is important because the stereocilia on the hair cells are polarized: they only give an impulse when moved in one defined direction.

The maximum decharge rate of a neuron is limited to about 1000 spikes/s and as a consequence the synchronization declines smoothly with increasing frequency, becoming negligible above 4–5 kHz. The role of the phase-locked activity is important for the sensation of musical pitch and timbre. Synchronized responses are used for the coding of the low pitch (as shown in VAM) and may play a role in the perception of enlarged octaves. Neural synchronization also improves absolute threshold sensitivity and sharpens spectral representation by synchronization and suppression. The sharpening plays an important role in timbre recognition where the strong components of the acoustic waveform are contained in the formant peaks.

B.3.3 Intensity

Intensity is coded by an increase in mean neuronal decharges. The dynamic range of an individual auditory neuron, however, is limited to about 20–30 dB, so that the total dynamic range of the human auditory system (about 140 dB) must be explained by taking into consideration more than one neuron. One hair cell with an optimal response to some particular frequency will indeed be excited by a range of lower and higher frequencies. In other words, one tone will stimulate an array of neurons. In that sense, the excitation pattern will reflect the transversal wave in the cochlear partition.

B.4 The Brain Stem and Cortex

The neuronal activity in the auditory nerve is the first of a chain of processing activities. The most important nuclei of the brainstem are the Cochlear Nucleus, the Lateral Lemniscus, the Inferior Colliculus, the Medial Geniculate Body, and the Auditory Cortex. Their specific function and mutual connections are beyond the scope of this book and it may suffice here to summarize some main characteristics of the auditory information processing at this level:

- **Feature Detection**. The neurons of the brain are specialized: they detect features in the signal and send the results to other neurons which detect other features.
- **Ordered**. Neuronal functions seem to be ordered. Tonotopy is one such ordering which shows that the cochlear frequency analysis is somehow reflected at higher centers. Tonotopy has been discovered at all major nuclei.
- **Hierarchic**. The neuronal functions are organized in a hierarchical way in that results of lower levels are further processed at higher levels.
- **Parallel**. The information processing is parallel. This feature explains why relatively slow processing units can lead to fast and intelligent reactions on complex stimuli.
- **Temporal Resolution**. Going more toward the center of the auditory system, the temporal resolution of the neurons is less fine. One may assume that larger auditory streams are processed at higher levels.

C. Normalization and Similarity Measures

This appendix contains some additional information about normalization and similarity measures. As shown in Fig. 4.1, the cognition module takes its input from the perception module. In practice, some aspects of normalization should be taken into consideration. Normalization depends on the properties of the data, as well as on the (di)similarity measures in the cognition module. In general, when the patterns are normalized beforehand, they should not be normalized during the matching process in SOM (6.2). For computational efficiency, it is advisable to adopt this strategy.

C.1 Similarity Measures

Candidates for similarity measures have been mentioned in the pattern recognition literature. Below we discuss the similarity measures that have been relevant for the present study.

The first measure is the *correlation coefficient* (cor) which is related to the *direction cosine* (dircos). The correlation coefficient is computational intensive but is an interesting measure to obtain an idea of the relationships between patterns. It is used for comparison of the model's structure with psychological data.

$$\text{cor} = \frac{\sum_k (x_k - \mu_x)(y_k - \mu_y)}{\sqrt{\sum_k (x_k - \mu_x)^2}\,\sqrt{\sum_k (y_k - \mu_y)^2}} \tag{C.1}$$

with $\mu_x = \frac{1}{k}\sum_k x_k$ and $\mu_y = \frac{1}{k}\sum_k y_k$

$$\text{dircos} = \frac{\sum_k x_k y_k}{\sqrt{\sum_k x_k^2}\sqrt{\sum_k y_k^2}}. \tag{C.2}$$

The *unnormalized correlation* (*ucor*) is equivalent to the inner or scalar product of two vectors. It is also related to the autocorrelation function. The unnormalized correlation is the most economical similarity measure from a computational point of view but if there are many zeros in the patterns, this measure might corrupt the comparison of the similarity, because the zero values in either x or y do not contribute to the measure of match.

$$\text{ucor} = \sum_k x_k y_k. \tag{C.3}$$

For that reason the distance as a measure of dissimilarity is more useful. The *Euclidian distance* between two patterns is given by

$$\text{eudist} = \sqrt{\sum_k (x_k - y_k)^2}. \tag{C.4}$$

The value of eudist is zero when both patterns match. Equation (C.4) is often used when the length of the patterns does not differ too much.

C.2 Towards a Psychoacoustic-Based Similarity Measure

An alternative approach, proposed by R. Parncutt, starts from the idea that measurements of similarity between tone images should somehow take into account the salience of tones. The similarity measure, called *commonality*, is expressed by

$$\text{comm} = \frac{\sum_k \sqrt{x_k y_k}}{\sqrt{\sum_k x_k \sum_k y_k}}. \tag{C.5}$$

Commonality is not used as a similarity measure in our study. According to Parncutt, however, it has the advantage of being intuitively appealing in the context of psychoacoustical considerations. He defines *tone salience* as the probability of being noticed or experienced. As a function of the residue weight it can be used to estimate the number of tones that is perceived simultaneously. According to this definition, the sum of the saliencies of all tone components in a chord is equal to the number of tones perceived. This is equal to 1 divided by the weight of the most salient tone component of the completion pattern.

This definition allows a transformation of a completion pattern (Sect. 5.3) into a salience pattern by means of[1]

$$O_i = \frac{1}{\sqrt{\sum_j \frac{R_j}{R_{\max}}}} * \frac{R_i}{R_{\max}}. \tag{C.6}$$

where O_i is the salience of component i, R_i the residue weight of component i, and $R_{\max}$ the highest residue weight. The introduction of the square root is based on psychoacoustical experiments described by Parncutt. It is possible, however, to relate this formula to aspects of synthetic and analytic listening.

From an alternative point of view, one may conceive of the completion pattern as a probability distribution for gestalt perception. R is then a discrete random variable that maps tone components onto scalar values. One

[1] The "O" is adopted from the Dutch term "opvallendheid" (=salience).

could define the probability of perceiving tone i as the R-weight of i given the total weight of the R-pattern:

$$P(R = R_i) = \frac{R_i}{\sum_j R_j}, \tag{C.7}$$

where $P(R = R_i)$ is the probability that the outcome of R is R_i. The sum of these probabilities is 1. This interpretation, however, suggests that all components can be perceived, although the perception of $R_{\max}$ is more probable than the others. But in the normal listening situation (synthetic listening), we do hear the residue pitch and it is only in the analytic mode that this interpretation could make any sense.

A more appealing point of view is therefore based on a ratio between both analytic and synthetic listening. Let us therefore start from the idea that the pitch with the highest weight is always the one perceived. Then, instead of speaking in terms of probabilities, it makes more sense to talk of *pregnance.* If $R_{\max}$ is the tone with the highest weight, then $P_{\max}$, its pregnance, should be equal to 1. The pregnance of the other tones in the R-pattern can then be defined with respect to the maximum, as in

$$P_i = \frac{R_i}{R_{\max}}. \tag{C.8}$$

Pregnance is a magnitude connected to the synthetic listening modality and the corresponding notion of the analytical listening modality is called *multiplicity.*

Multiplicity provides an estimate of the number of tones perceived and may be defined as the sum of all pregnancies, as in

$$M = \sum_j P_j = \frac{\sum_j R_j}{R_{\max}}, \tag{C.9}$$

where M is the multiplicity of the entire auditory image. This can be related to Parncutt's observation that the square root of this sum gives a more realistic approximation:

$$M' = \sqrt{M} = \sqrt{\sum_j P_j} = \frac{\sqrt{\sum_j R_j}}{\sqrt{R_{\max}}}. \tag{C.10}$$

If we now return to (C.6), we see that the salience of a component is defined as the ratio of the pregnance of the component and the multiplicity (M') of the component's auditory image. In more general terms: salience expresses the ratio of synthetic and analytical listening.

$O_{\max}$, the component of the O-pattern with the highest weight, is equal to 1 divided by M'. Consequently, if we know the component with the maximum salience it is easy to deduce the multiplicity from it:

$$M' = \frac{1}{O_{\max}}. \tag{C.11}$$

The application of (C.6) to R-patterns introduces a normalization that has been proven to be useful (at least for 12-dimensional patterns).

References

Chapter 1

1.1 J.M. Grey: Multidimensional perceptual scaling of musical timbres. *J. Acoust. Soc. Am.* **61**, 1270–1277 (1977)

1.2 J.M. Grey: Timbre discrimination in musical patterns. *J. Acoust. Soc. Am.* **64**, 467–472 (1978)

1.3 J.M. Grey, J.W. Gordon: Perceptual effects of spectral modifications on musical timbres. *J. Acoust. Soc. Am.* **63**, 1493–1500 (1978).

1.4 R. Kendall, E. Carterette: Perceptual scaling of simultaneous wind instrument timbres. *Music Perception* **8**, 369–404 (1991)

1.5 C. Krumhansl: *Cognitive Foundations of Musical Pitch* (Oxford Univ. Press, New York 1990)

1.6 U. Seifert: The schema concept - a critical review of its development and current use in cognitive science and research on music perception. In *IX Colloquium on Musical Informatics*, ed. by A. Camuri, C. Canepa (AIMI/DIST, Genova 1991)

1.7 R.N. Shepard: Structural representations of musical pitch. In *The Psychology of Music*, ed. by D. Deutsch (Academic, New York 1982)

1.8 R.N. Shepard, S. Chipman: Second-order isomorphism of internal representations - shapes of states. *Cognitive Psychology* **1**, 1–17 (1970)

1.9 K. Ueda, K. Ohgushi: Perceptual components of pitch - spatial representation using a multidimensional scaling technique. *J. Acoust. Soc. Am.* **82**, 1193–1203 (1987)

1.10 D. Wessel: Timbre space as a musical control structure. *Computer Music J.* **3**, 45–52 (1979)

Chapter 2

2.1 E. Clarke: Categorical rhythm perception - an ecological perspective. In *Action and Perception in Rhythm and Music*, ed. by A. Gabrielsson (The Royal Swedish Academy of Music, Stockholm 1987)

2.2 C. Dahlhaus: *Untersuchungen über die Entstehung der harmonischen Tonalität (Studies on the Origin of Harmonic Tonality, transl. by R. O. Gjerdingen)* (Princeton Univ. Press, Princeton, NJ 1966/1990)

2.3 A. Gabrielsson: Once again: the theme from Mozart's piano sonata in A major (KV 331) - a comparision of five performances. In *Action and Perception in Rhythm and Music*, ed. by A. Gabrielsson (The Royal Swedish Academy of Music, Stockholm 1987)

2.4 A. Gabrielsson: Timing in music performance and its relations to music experience. In *Generative Processes in Music*, ed. by J. A. Sloboda (Clarendon Press, Oxford 1988)

2.5 W.M. Hartmann: On the origin of the enlarged melodic octave. *J. Acoust. S. Am.* **93**, 3400–3409 (1993)

2.6 H. Helmholtz: *Die Lehre von den Tonempfindungen als physiologische Grundlage für die Theorie der Musik.* (Georg Olms, Hildesheim 1863/1968)

2.7 A. Kameoka, M. Kuriyagawa: Consonance theory Part I - consonance of dyads. *J. Acoust. Soc. Am.* **45**, 1451–1459 (1969)

2.8 C. Krumhansl: Tonal and harmonic hierarchies. In *Harmony and Tonality*, ed. by J. Sundberg (Royal Swedish Academy of Music, Stockholm 1987)

2.9 C. Krumhansl: *Cognitive Foundations of Musical Pitch* (Oxford Univ. Press, New York 1990)

2.10 C. Krumhansl, E. Kessler: Tracing the dynamic changes in perceived tonal organization in a spatial representation of musical keys. *Psychological Review* **89**, 334–368 (1982)

2.11 C. Krumhansl, R.N. Shepard: Quantification of the hierarchy of tonal functions within a diatonic context. *J. of Experimental Psychology - Human Perception and Performance*, 5:579–594, 1979.

2.12 J.B. Kruskal, M. Wish: *Multidimensional Scaling.* (Sage Publ., Beverly Hills, CA 1978)

2.13 E. Kurth: *Die Voraussetzungen der Theoretischen Harmonik und der tonalen Darstellungssysteme.* (Musikverlag Emil Katzbichler, München 1913/1973)

2.14 F. Lerdahl: Tonal pitch space. *Music Perception* **5**, 315–350 (1988)

2.15 H.J. Maxwell: An expert system for harmonic analysis of tonal music. In *Understanding Music with AI - Perspectives on Music Cognition*, ed. by M. Balaban, K. Ebcioglu, O. Laske (MIT Press, Cambridge, MA 1992)

2.16 G. Mazzola, H.G. Wieser, V. Brunner, D. Muzzulini: A symmetry-oriented mathematical model of classical counterpoint and related neurophysiological investigations by depth EEG. *Computers Math. Applic.* **17**, 539–594 (1989)

2.17 B.C.J. Moore, B.R. Glasberg: Suggested formulae for calculating auditory-filter bandwidths and excitation patterns. *J. Acoust. Soc. Am.* **74**, 750–753 (1983)

2.18 J.P. Rameau: *Traité de l'Harmonie* (Broude Brothers, New York 1722/1965)

2.19 B. Repp: Patterns of expressive timing in performances of a Beethoven minuet by 19 famous pianists. *J. Acoust. Soc. Am.* **88**, 622–641 (1990)

2.20 L.H. Schaffer, N.P. Todd: The interpretive component in musical performance. In *Action and Perception in Rhythm and Music*, ed. by A. Gabrielsson (The Royal Swedish Academy of Music, Stockholm 1987)

2.21 A. Schönberg: *Harmonielehre* (Universal Edition, Heidelberg 1922/1986)

2.22 R.N. Shepard: Structural representations of musical pitch. In *The Psychology of Music*, ed. by D. Deutsch (Academic, New York 1982)

2.23 C. Stumpf: *Tonpsychology I* (Hirzel, Leipzig 1883)

2.24 C. Stumpf: *Tonpsychology II* (Hirzel, Leipzig 1890)

2.25 J. Sundberg, A. Friberg, L. Frydén: Common secrets of musicians and listeners - an analysis-by-synthesis study of musical performance. In *Representing Musical Structure*, ed. by P. Howell, R. West, I. Cross (Academic, London 1991)

2.26 J. Sundberg, L. Fryden: Melodic charge and music performance. In *Harmony and Tonality*, ed. by J. Sundberg (Royal Swedish Academy of Music, Stockholm 1987)

2.27 E. Terhardt: The concept of musical consonance - a link between music and psychoacoustics. *Music Perception* **1**, 276–295 (1984)

2.28 N. Todd: The dynamics of dynamics - a model of musical expression. *J. Acoust. Soc. Am.* **91**, 3540–3550 (1992)

2.29 R.M. Warren: Helmholtz and his continuing influence. *Music Perception* **1**, 253–275 (1984)

2.30 A. Wellek: *Musikpsychologie und Musikästhetik - Grundriss der Systematischen Musikwissenschaft* (Akademische Verlagsgesellschaft, Frankfurt am Main 1963)

2.31 T. Wishart: From architecture to chemistry. *Interface - J. New Music Research* **22**, 301–315 (1993)

2.32 I. Xenakis: Towards a philosophy of music. *Gravesäner Blätter* **29**, 39–52 (1966)

2.33 E. Zwicker, H. Fastl: *Psychoacoustics - Facts and Models* (Springer Ser. Inf. Sci., Vol. 22, Berlin, Heidelberg 1990)

2.34 E. Zwicker, R. Feldtkeller: *Das Ohr als Nachrichtenempfänger* (Hirzel, Stuttgart 1967)

Chapter 3

3.1 E. Burns: Circularity in relative pitch judgments for inharmonic complex tones - the Shepard demonstration revisited again. *Percept. Psychophys.* **30**, 467–472 (1981)

3.2 D. Deutsch: A musical paradox. *Music Perception* **3**, 275–280 (1986)

3.3 D. Deutsch: The tritone paradox - an influence of language on music perception. *Music Perception* **8**, 335–347 (1991)

3.4 D. Deutsch, R.C. Boulanger: Octave equivalence and the immediate recall of pitch sequences. *Music Perception* **2**, 40–51 (1984)

3.5 D. Deutsch, W.L. Kuyper, Y. Fisher: The tritone paradox - its presence and form of distribution in a general population. *Music Perception* **5**, 79–92 (1987)

3.6 D. Deutsch, T. North, L. Ray: The tritone paradox - correlate with the listener's vocal range for speech. *Music Perception* **7**, 371–384 (1990)

3.7 A. Forte: *The Structure of Atonal Music* (Yale Univ. Press, New Haven, CT 1973)

3.8 M. Leman: Symbolic and subsymbolic description of music. In *Music Processing*, ed. by G. Haus . (A-R Editons, Madison, Wisconsin 1993)

3.9 M.V. Mathews: *The Technology of Computer Music* (MIT Press, Cambridge, MA 1969)

3.10 M.V. Mathews, R. Pierce, A. Reeves, L.A. Roberts: Theoretical and experimental explorations of the Bohlen-Pierce scale. *J. Acoust. Soc. Am.* **84**, 1214–1222 (1988)

3.11 J. Nakajima, H. Minami, T. Tsumura, H. Kunisaki, S. Ohnishi, R. Teranishi: Dynamic pitch perception for complex tones of periodic spectral patterns. *Music Perception* **8**, 291–314 (1991)

3.12 Y. Nakajima, T. Tsumura, S. Matsuura, H. Minami, R. Teranishi: Dynamic pitch perception for complex tones derived from major triads. *Music Perception* **6**, 1–20 (1988)

3.13 G. Révész: *Inleiding tot de Muziekpsychologie* (N.V. Noord-Hollandsche Uitgevers Maatschappij, Amsterdam 1944)

3.14 J. Risset: Hauteur et timbre des sons. Technical Report IRCAM Nr. 11 (Centre Georges Pompidou, Paris 1978)

3.15 R.N. Shepard: Circularity in judgments of relative pitch. *J. Acoust. Soc. Am.* **36**, 2346–2353 (1964)
3.16 K. Ueda, K. Ohgushi: Perceptual components of pitch - spatial representation using a multidimensional scaling technique. *J. Acoust. Soc. Am.* **82**, 1193–1203 (1987)
3.17 L. van Noorden: Two channel pitch perception. In *Music, Mind and Brain - the Neuropsychology of Music*, ed. by M. Clynes (Plenum Press, London 1982)
3.18 G. von Bismarck: Sharpness as an attribute of the timbre of steady sounds. *Acustica* **30**, 159–172 (1974)
3.19 G. von Bismarck: Timbre of steady sounds - a factorial investigation of its verbal attributes. *Acustica* **30**, 146–159 (1974)
3.20 E. Zwicker, H. Fastl: *Psychoacoustics - Facts and Models* (Springer Ser. Inf. Sci., Vol. 22, Berlin, Heidelberg 1990)

Chapter 4

4.1 A.S. Bregman: *Auditory Scene Analysis - the Perceptual Organization of Sound* (MIT Press, Cambridge, MA 1990)
4.2 G. Brown: Computational auditory scene analysis. Technical report (Dept. of Comp. Sc., Univ. of Sheffield 1992)
4.3 P. Cosi, G. De Poli, G. Lauzzana: Auditory modelling and self-organizing neural networks for timbre classification. *J. New Music Research* **23**, 71–98 (1994)
4.4 P. Dallos: Cochlear neurobiology - some key experiments and concepts of the past two decades. In *Auditory Function - Neurobiological Bases of Hearing*, ed. by G.M. Edelman, W. Gall, W. Cowan (Wiley, New York 1988)
4.5 J.D. Durrant, J.H. Lovrinic: *Bases of Hearing Science* (Williams and Wilkins, Baltimore 1984)
4.6 G.M. Edelman, W. Gall, W. Cowan (eds.): *Auditory Function - Neurobiological Bases of Hearing* (Wiley, New York 1988)
4.7 E. Javel, J. McGee, J.W. Horst, G.R. Farley: Temporal mechanisms in auditory stimulus coding. In *Auditory Function - Neurobiological Bases of Hearing*, ed. by G.M. Edelman, W. Gall, W. Cowan (Wiley, New York 1988)
4.8 M.V. Mathews: *The Technology of Computer Music* (MIT Press, Cambridge, MA 1969)
4.9 J.O. Pickles: *An introduction to the physiology of hearing* (Academic, London 1988)
4.10 C.E. Schreiner, G. Langner: Coding of temporal patterns in the central auditory nervous system. In *Auditory Function - Neurobiological Bases of Hearing*, ed. by G.M. Edelman, W. Gall, W. Cowan (Wiley, New York 1988)
4.11 C.E. Schreiner, M.M. Merzenich: Elements of signal coding in the auditory nervous system. In *Organization of Neural Networks - Structures and Models*, ed. by W. von Seelen, G. Shaw, U.M. Leinhos (VCH, Weinheim 1988)
4.12 N. Suga: Auditory neuro–ethology and speech processing - complex sound processing by combination–sensitive neurons. In *Auditory Function - Neurobiological Bases of Hearing*, ed. by G.M. Edelman, W. Gall, W. Cowan (Wiley, New York 1988)
4.13 N. Todd: The auditory "primal sketch" - a multiscale model of rhythmic grouping. *J. New Music Research* **23**, 25–70 (1994)

4.14 B. Vercoe: CSOUND - a manual for the audio processing system and supporting programs. Technical Report (Media Lab MIT, Cambridge, MA 1986)
4.15 E.D. Young, W.P. Shofner, J.A. White, J.M. Robert, H.F. Voigt: Response properties of cochlear nucleus neurons in relationship to physiological mechanisms. In *Auditory Function - Neurobiological Bases of Hearing*, ed. by G.M. Edelman, W. Gall, W. Cowan (Wiley, New York 1988)
4.16 S. Zeki: *A Vision of the Brain* (Blackwell, Oxford 1993)

Chapter 5

5.1 P. Assmann, Q. Summerfield: Modeling the perception of concurrent vowels - vowels with different fundamental frequencies. *J. Acoust. Soc. Am.* **88**, 680–697 (1990)
5.2 S. A. Gelfand: *Hearing - an Introduction to Psychological and Physiological Acoustics* (Marcel Dekker, New York 1981)
5.3 W. Hess: *Pitch Determination of Speech Signals - Algorithms and Devices* (Springer Ser. Inf. Sc., Vol. 3, Berlin, Heidelberg 1983)
5.4 L. Van Immerseel: *Een Functioneel Gehoormodel voor de Analyse van Spraak bij Spraakherkenning.* PhD thesis (Univ. of Ghent, Ghent 1993)
5.5 L. Van Immerseel, J.P. Martens: Pitch and voiced/unvoiced determination with an auditory model. *J. Acoust. Soc. Am.* **91**, 3511–3526 (1992)
5.6 E. Javel, J. McGee, J.W. Horst, G.R. Farley: Temporal mechanisms in auditory stimulus coding. In *Auditory Function - Neurobiological Bases of Hearing*, ed. by G.M. Edelman, W. Gall, W. Cowan (Wiley, New York 1988)
5.7 M. Leman: Symbolic and subsymbolic information processing in models of musical communication and cognition. *Interface - J. New Music Research* **18**, 141–160 (1989)
5.8 M. Leman: Emergent properties of tonality functions by self-organization. *Interface - J. New Music Research* **19**, 85–106 (1990)
5.9 M. Leman: Künstliche Neuronale Netzwerke - Neue Ansätze zur ganzheitlichen Informations-verarbeitung in der Musikforschung. In *Computer in der Musik*, ed. by H. Schaffrath (J.B. Metzler, Stuttgart 1991)
5.10 M. Leman: The ontogenesis of tonal semantics - results of a computer study. In *Music and Connectionism*, ed. by P. Todd, G. Loy (MIT Press, Cambridge, MA 1991)
5.11 M. Leman, P. Van Renterghem: Transputer implementation of the Kohonen feature map for a music recognition task. In *Proc. of the Second International Transputer Conf.: Transputers for Industrial Applications II* (BIRA, Belgian Institute for Automatic Control, Antwerpen 1989)
5.12 R. Meddis, M.J. Hewitt: Virtual pitch and phase sensitivity of a computer-model of the auditory periphery I - pitch identification. *J. Acoust. Soc. Am.* **89**, 2866-2894 (1991)
5.13 R. Parncutt: Revision of Terhardt's psychoacoustical model of the roots of a musical chord. *Music perception* **6**, 65–94 (1988)
5.14 R. Parncutt: *Harmony - a Psychoacoustical Approach* (Springer Ser. Inf. Sc., Vol.19, Berlin, Heidelberg 1989)
5.15 R. Parncutt: Template-matching models of musical pitch and rhythm perception. *J. New Music Research* **23**, 145–167 (1994)
5.16 E. Terhardt: Pitch, consonance, and harmony. *J. Acoust. Soc. Am.* **55**, 1061–1069 (1974)

5.17 E. Terhardt, G. Stoll, M. Seewann: Algorithm for extraction of pitch and pitch salience from complex tonal signals. *J. Acoust. Soc. Am.* **71**, 679–688 (1982)
5.18 E. Terhardt, G. Stoll, M. Seewann: Pitch of complex signals according to virtual-pitch theory - tests, examples, and predictions. *J. Acoust. Soc. Am.* **71**, 671–628 (1982)
5.19 L. van Noorden: Two channel pitch perception. In *Music, Mind and Brain - the Neuropsychology of Music*, ed. by M. Clynes (Plenum Press, London 1982)

Chapter 6

6.1 J.P. Changeux: *L' homme neuronal* (Fayard, Paris 1983)
6.2 G.M. Edelman: *Neural Darwinism - the Theory of Neuronal Group Selection* (Basic Books, New York 1987)
6.3 G.M. Edelman, W. Gall, W. Cowan (eds.): *Auditory Function - Neurobiological Bases of Hearing* (Wiley, New York 1988)
6.4 F.J. Fétis: *Traité Complet de la Théorie et de la Pratique de l'Harmonie* (Maurice Schlesinger, Paris 1844)
6.5 H. Haken: Synergetics as a tool for the conceptualization and mathematization of cognition and behaviour - how far can we go. In *Synergetics of Cognition*, ed. by H. Haken, M. Stadler (Springer Ser. Synergetics, Berlin, Heidelberg 1990)
6.6 H. Haken, M. Stadler (eds.): *Synergetics of Cognition* (Springer Ser. Synergetics, Berlin, Heidelberg 1990)
6.7 J. Hopfield: Neural networks and physical systems with emergent collective computational abilities. *Proc. N.A.S.USA 79*, 2554–2558 (1982)
6.8 Y. Kamp, M. Hasler: *Recursive Neural Networks for Associative Memory* (Wiley, Chichester, UK 1990)
6.9 G. Kanizsa, R. Luccio: The phenomenology of autonomous order formation in perception. In *Synergetics of Cognition*, ed. by H. Haken, M. Stadler (Springer Ser. Synergetics, Berlin, Heidelberg 1990)
6.10 J. A.S. Kelso: Phase transitions - foundations of behavior. In *Synergetics of Cognition*, ed. by H. Haken, M. Stadler (Springer Ser. Synergetics, Berlin, Heidelberg 1990)
6.11 T. Kohonen: *Self-Organization and Associative Memory* (Springer Ser. Inf. Sc., Vol.8, Berlin, Heidelberg 1984)
6.12 T. Kohonen: The self-organizing map. *IEEE Proc.* **78**, 1464–1480 (1990)
6.13 M. Leman: *Een Model van Toonsemantiek - naar een Theorie en Discipline van de Muzikale Verbeelding.* PhD thesis (Univ. of Ghent, Ghent 1991)
6.14 M. Leman: Complex dynamics in music cognition - aspects of tone center perception. In *Proc. of the 13th Int. Conf. on Cybernetics* (International Association for Cybernetics, Namur 1993)
6.15 M. Leman: Tone center attraction dynamics - an approach to schema-based tone center recognition of musical signals. In *Atti di X Colloquio di Informatica Musicale* (AIMI/LIM, Milano 1993)
6.16 M. Leman: Schema-based tone center recognition of musical signals. *J. New Music Research* **23**, 169–204 (1994)
6.17 M. Leman, P. Van Renterghem: Transputer implementation of the Kohonen feature map for a music recognition task. In *Proc. of the Second International Transputer Conf.: Transputers for Industrial Applications II* (BIRA, Belgian Institute for Automatic Control, Antwerpen 1989)

6.18 H.R. Maturana, F.J. Varela: *De Boom der Kennis - Hoe Wij de Wereld door onze Eigen Waarneming Creëren* (Uitgeverij Contact, Amsterdam 1984)

6.19 P. Morasso, V. Sanguineti: Self-organizing topographic maps and motor planning. Technical report (Univ. di Genova, D.I.S.T., Genova 1994)

6.20 R. Serra, G. Zanarini: *Complex Systems and Cognitive Processes.* (Springer, Berlin, Heidelberg 1990)

6.21 M. Stadler, P. Kruse: The self-organization perspective in cognition research - historical remarks and new experimental approaches. In *Synergetics of Cognition*, ed. by H. Haken and M. Stadler (Springer Ser. in Synergetics, Berlin, Heidelberg 1990)

6.22 L. Steels: Cooperation between distributed agents through self-organization. Technical Report AI-memo 89-5 (VUB-AI Lab, Brussel 1989)

6.23 A.C. Zimmer: Autonomous organization in perception and motor control. In *Synergetics of Cognition*, ed. by H. Haken and M. Stadler (Springer Ser. in Synergetics, Berlin, Heidelberg 1990)

Chapter 7

7.1 S.C. Ahalt, A.K. Krishnamurthy, P. Chen, D.E. Melton: Competitive learning algorithms for vector quantization. *Neural Networks* **3**, 277–290 (1990)

7.2 H. Bruhn: *Harmonielehre als Grammatik der Musik* (Psychologie Verlags Union, München 1988)

7.3 T. Kohonen: The self-organizing map. *IEEE Proc.* **78**, 1464–1480 (1990)

7.4 C. Krumhansl: *Cognitive Foundations of Musical Pitch* (Oxford Univ. Press, New York 1990)

7.5 M. Leman: Emergent properties of tonality functions by self-organization. *Interface - J. New Music Research* **19**, 85–106 (1990)

7.6 M. Leman: *Een Model van Toonsemantiek - naar een Theorie en Discipline van de Muzikale Verbeelding.* PhD thesis (Univ. of Ghent, Ghent 1991)

7.7 M. Leman: The ontogenesis of tonal semantics - results of a computer study. In *Music and Connectionism*, ed. by P. Todd, G. Loy (MIT Press, Cambridge, MA 1991)

7.8 M. Leman: Tone context and the complex dynamics of tone semantics. In *Proc. of the KlangArt Kongress*, ed. by B. Enders (Schott's Sohne, Mains 1991)

7.9 R. Parncutt: *Harmony - a Psychoacoustical Approach* (Springer Ser. Inf. Sc., Vol.19, Berlin, Heidelberg 1989)

7.10 B. Vercoe: CSOUND—a manual for the audio processing system and supporting programs. Technical Report (Media Lab MIT, Cambridge, MA 1986)

Chapter 8

8.1 D. Butler, H. Brown: Describing the mental representation of tonality in music. In *Musical Perceptions*, ed. by R. Aiello, J.A. Sloboda (Oxford Univ.Press, New York 1994)

8.2 G.J. Chappell, J.G. Taylor: The temporal Kohonen map. *Neural Networks* **6**, 441–445 (1993)

8.3 F.J. Fétis: *Traité Complet de la Théorie et de la Pratique de l'Harmonie.* (Maurice Schlesinger, Paris 1844)

8.4 P. Fraisse: Rhythm and tempo. In *The Psychology of Music*, ed. by D. Deutsch (Academic Press, New York 1982)

8.5 L. Knopoff, W. Hutchinson: An index of melodic activity. *Interface - J. New Music Research* **7**, 205–229 (1978)

8.6 M. Naranjo: Apprentissage et reconnaissance de séquences musicales. In *Proc. of the 13th Int. Conf. on Cybernetics* (International Association for Cybernetics, Namur 1993)

8.7 R. Plomp: The role of modulation in hearing. In *Hearing - physiological bases and psychophysics*, ed. by R. Klinke, R. Hartmann (Springer, Berlin, Heidelberg 1983)

8.8 E. Pöppel: The measurement of music and the cerebral clock - a new theory. *Leonardo* **22**, 83–89 (1989)

8.9 R.A. Rasch: The perception of simultaneous notes such as in polyphonic music. *Acustica* **40**, 21–33 (1978)

8.10 R.A. Rasch: Synchronization in performed ensemble music. *Acustica* **43**, 121–131 (1979)

8.11 J. Vos, R. Rasch: The perceptual onset of musical tones. In *Music, Mind and Brain - the Neuropsychology of Music*, ed. by M. Clynes (Plenum Press, London 1982)

Chapter 9

9.1 D. J. Amit: *Modeling Brain Function - the World of Attractor Neural Networks* (Cambridge Univ. Press, Cambridge, MA 1989)

9.2 H. Haken, M. Stadler (eds.): *Synergetics of Cognition* (Springer Ser. Synergetics, Berlin, Heidelberg 1990)

9.3 M. Leman: The theory of tone semantics - concept, foundation, and application. *Minds and Machines* **2**, 345–363 (1992)

Chapter 10

10.1 J. Bharucha: Music cognition and perceptual facilitation - a connectionist framework. *Music Perception* **5**, 1–30 (1987)

10.2 J. Bharucha, P. Todd: Modeling the perception of tonal structure with neural nets. In *Music and Connectionism*, ed. by P. Todd, G. Loy (MIT Press, Cambridge, MA 1991)

10.3 G.A. Carpenter, S. Grossberg: The art of adaptive pattern recognition by a self-organizing neural network. *IEEE-Computer* **21**, 77–88 (1988)

10.4 R. Eberlein, J.P. Fricke: *Kadenzwahrnehmung und Kadenzgeschichte - ein Beitrag zu einer Grammatik der Musik* (Peter Lang, Frankfurt am Main 1992)

10.5 R.O. Gjerdingen: Categorization of music patterns by self-organizing neuronlike networks. *Music Perception* **7**, 339–369 (1990)

10.6 R.O. Gjerdingen: Learning syntactically significant temporal patterns of chords - a masking field embedded in an ART 3 architecture. *Neural Networks* **5**, 551–564 (1992)

10.7 S.R. Holtzman: A program for key determination. *Interface - J. New Music Research* **6**, 29–56 (1977)

10.8 C. Longuet-Higgins: The three dimensions of harmony. In *Mental Processes - Studies in Cognitive Science*, ed. by C. Longuet-Higgins (MIT Press, Cambridge, MA 1962/1987)
10.9 C. Longuet-Higgins: The grammar of music. In *Mental Processes - Studies in Cognitive Science*, ed. by C. Longuet-Higgins (MIT Press, Cambridge, MA 1978/1987)
10.10 C. Longuet-Higgins, M.J. Steedman: On interpreting bach. In *Mental Processes - Studies in Cognitive Science*, ed. by C. Longuet-Higgins (MIT Press, Cambridge, MA 1971/1987)
10.11 H.J. Maxwell: An expert system for harmonic analysis of tonal music. In *Understanding Music with AI - Perspectives on Music Cognition*, ed. by M. Balaban, K. Ebcioglu, O. Laske (MIT Press, Cambridge, MA 1992)
10.12 D.J. Povel, R.Van Egmond: The function of accompanying chords in the recognition of melodic fragments. *Music Perception* **11**, 101–115 (1993)
10.13 P. Todd: A connectionist approach to algorithmic composition. In *Music and Connectionism*, ed. by P. Todd, G. Loy (MIT Press, Cambridge, MA 1991)

Chapter 11

11.1 G. Brown: Computational auditory scene analysis. Technical report (Dept. of Comp. Sc., Univ. of Sheffield 1992)
11.2 J.C. Brown: Determination of the meter of musical scores by autocorrelation. *J. Acoust. Soc. Am.* **94**, 1953–1957 (1993)
11.3 E. Clarke: Categorical rhythm perception - an ecological perspective. In *Action and Perception in Rhythm and Music*, ed. by A. Gabrielsson (The Royal Swedish Academy of Music, Stockholm 1987)
11.4 E. Clarke: Generative principles in music performance. In *Generative Processes in Music*, ed. by J. A. Sloboda (Clarendon Press, Oxford 1988)
11.5 M. Clynes, J. Walker: Music as time's measure. *Music Perception* **4**, 85–120 (1986)
11.6 P. Cosi, G.De Poli, G. Lauzzana: Auditory modelling and self-organizing neural networks for timbre classification. *J. New Music Research* **23**, 71–98 (1994)
11.7 P. Desain, H. Honing: *Music, Mind and Machine - Studies in Computer Music, Music Cognition and Artificial Intelligence* (Thesis publishers, Amsterdam 1992)
11.8 P. Fraisse: Rhythm and tempo. In *The Psychology of Music*, ed. by D. Deutsch (Academic, New York 1982)
11.9 A. Gabrielsson: Once again - the theme from Mozart's piano sonata in A major (KV 331) - a comparision of five performances. In *Action and Perception in Rhythm and Music*, ed. by A. Gabrielsson (The Royal Swedish Academy of Music, Stockholm 1987)
11.10 J.W. Gordon, J.M. Grey: Perception of spectral modifications on orchestral instrument tones. *Computer Music J.* **2**, 24–31 (1978)
11.11 J.M. Grey: Multidimensional perceptual scaling of musical timbres. *J. Acoust. Soc. Am.* **61**, 1270–1277 (1977)
11.12 J.M. Grey: Timbre discrimination in musical patterns. *J. Acoust. Soc. Am.* **64**, 467–472 (1978)
11.13 J.M. Grey, J.W. Gordon: Perceptual effects of spectral modifications on musical timbres. *J. Acoust. Soc. Am.* **63**, 1493–1500 (1978)

11.14 M.R. Jones: Perspectives on musical time. In *Action and Perception in Rhythm and Music*, ed. by A. Gabrielsson (The Royal Swedish Academy of Music, Stockholm 1987)
11.15 C.S. Lee: The perception of metrical structure - experimental evidence and a model. In *Representing Musical Structure*, ed. by P. Howell, R. West, I. Cross (Academic, London 1991)
11.16 M. Leman (ed.): Auditory models in music research (I). Special issue of J. New Music Research (Swets & Zeitlinger, Lisse 1994)
11.17 M. Leman (ed.): Auditory models in music research (II). Special issue of J. New Music Research (Swets & Zeitlinger, Lisse 1994)
11.18 D. Mellinger: *Event Formation and Separation in Musical Sound.* PhD thesis (Stanford University, Stanford 1991)
11.19 B.O. Miller, D.L. Scarborough, J.A. Jones: On the perception of meter. In *Understanding Music with AI - Perspectives on Music Cognition*, ed. by M. Balaban, K. Ebcioglu, O. Laske (MIT Press, Cambridge, MA 1992)
11.20 B. Repp: Patterns of expressive timing in performances of a Beethoven minuet by 19 famous pianists. *J. Acoust. Soc. Am.* **88**, 622–641 (1990)
11.21 P. Schaeffer: *Traité des Objets Musicaux - Essai Interdisciplines* (Editions du Seuil, Paris 1966)
11.22 L.H. Schaffer, N.P. Todd: The interpretive component in musical performance. In *Action and Perception in Rhythm and Music*, ed. by A. Gabrielsson (The Royal Swedish Academy of Music, Stockholm 1987)
11.23 S. Seneff: A joint synchrony/mean-rate model of auditory speech processing. *J.of Phonetics* **16**, 55–76 (1988)
11.24 L.S. Smith: Sound segmentation using onsets and offsets. *J. New Music Research* **23**, 11–23 (1994)
11.25 N. Todd: The dynamics of dynamics - a model of musical expression. *J. Acoust. Soc. Am.* **91**, 3540–3550 (1992)
11.26 N. Todd: The auditory "primal sketch" - a multiscale model of rhythmic grouping. *J. New Music Research* **23**, 25–70 (1994)
11.27 P. Toiviainen: The organisation of timbres - a two-stage neural network model. In *Proc. of the ECAI-92 workshop on AI and Music*, ed. by G. Widmer (Austrian Society for AI, Vienna 1992)
11.28 E.D. Young, W.P. Shofner, J.A. White, J.M. Robert, H.F. Voigt: Response properties of cochlear nucleus neurons in relationship to physiological mechanisms. In *Auditory Function - Neurobiological Bases of Hearing*, ed. by G.M. Edelman, W. Gall, W. Cowan (Wiley, New York 1988)

Chapter 12

12.1 L. Apostel, H. Sabbe, F. Vandamme (eds.): *Reason, Emotion and Music - Towards a Common Structure for Arts, Sciences and Philosophies, Based on a Conceptual Framework for the Description of Music* (Communication & Cognition, Ghent 1986)
12.2 D. Batens: Meaning, acceptance and dialectics. In *Change and Progress in Modern Science*, ed. by J. Pitt (Reidel, Dordrecht 1985)
12.3 D. Batens: Do we need a hierarchical model of science? In *Inference, Explanation, and Other Frustrations. Essays in the Philosophy of Science*, ed. by J. Earman (University of California Press, Oxford 1991)
12.4 J.L. Broeckx: *Muziek, Ratio en Affect - Over de Wisselwerking van Rationeel Denken en Affectief Beleven bij Voortbrengst en Ontvangst van Muziek.* (Metropolis, Antwerpen 1981)

12.5 E. de Boer: Pattern recognition of spectral information in hearing. In *Organization of Neural Networks - Structures and Models*, ed. by W. von Seelen, G. Shaw, U.M. Leinhos (VCH, Weinheim 1988)

12.6 J.D. Durrant, J.H. Lovrinic: *Bases of Hearing Science* (Williams and Wilkins, Baltimore 1984)

12.7 J. Fodor: *Representations.* (MIT Press, Cambridge, MA 1981)

12.8 J. Fodor: *The Modularity of Mind - an Essay on Faculty Psychology* (MIT Press, Cambridge, MA 1983)

12.9 J.J. Gibson: *The Perception of the Visual World* (Greenwood, Westport, Conn. 1950)

12.10 J.J. Gibson: *The Senses Considered as Perceptual Systems* (Houghton-Mifflin, Boston 1966)

12.11 H. Haken, M. Stadler (eds.): *Synergetics of Cognition* (Springer Ser. Synergetics, Berlin, Heidelberg, 1990)

12.12 A.J. Houtsma, J.L. Goldstein: The central origin of the pitch of complex tones - evidence from musical interval recognition. *J. Acoust. Soc. Am.* **51**, 520–529 (1972)

12.13 T. Kohonen: *Self-Organization and Associative Memory* (Springer Ser. Inf. Sc., Vol.8, Berlin, Heidelberg 1984)

12.14 G. Langner: Neuronal mechanisms for a periodicity analysis in the time domain. In *Hearing - Physiological Bases and Psychophysics*, ed. by R. Klinke, R. Hartmann (Springer, Berlin, Heidelberg 1983)

12.15 G. Langner, C.E. Schreiner: Periodicity coding in the inferior colliculus of the cat (I) - neuronal mechanisms. *J.of Neurophysiology* **60**, 1799–1822 (1988)

12.16 D. Marr: *Vision - a Computational Investigation into the Human Representation and Processing of Visual Information* (Freeman, San Francisco 1982)

12.17 M.V. Mathews, J.R. Pierce, L.A. Roberts: Harmony and new scales. In *Harmony and Tonality*, ed. by J. Sundberg (Royal Swedish Academy of Music, Stockholm 1987)

12.18 M.V. Mathews, R. Pierce, A. Reeves, L.A. Roberts: Theoretical and experimental explorations of the Bohlen-Pierce scale. *J. Acoust. Soc. Am.* **84**, 1214–1222 (1988)

12.19 I. Peretz, J. Morais: Music and modularity. In *Music and the Cognitive Sciences*, ed. by S. McAdams, I. Deliège (Harwood Academic, London 1989)

12.20 J.O. Pickles: *An Introduction to the Physiology of Hearing* (Academic, London 1988)

12.21 H.J. Reitboeck, R. Eckhorn, M. Arndt, P. Dicke: A model for feature linking via correlated neural activity. In *Synergetics of Cognition*, ed. by H. Haken, M. Stadler (Springer Ser. Synergetics, Berlin, Heidelberg 1990)

12.22 C.E. Schreiner, G. Langner: Coding of temporal patterns in the central auditory nervous system. In *Auditory Function - Neurobiological Bases of Hearing*, ed. by G.M. Edelman, W. Gall, W. Cowan (Wiley, New York 1988)

12.23 C.E. Schreiner, M.M. Merzenich: Elements of signal coding in the auditory nervous system. In *Organization of Neural Networks - Structures and Models*, ed. by W. von Seelen, G. Shaw, U.M. Leinhos (VCH, Weinheim 1988)

12.24 R.N. Shepard: Structural representations of musical pitch. In *The Psychology of Music*, ed. by D. Deutsch (Academic Press, New York 1982)

12.25 R.N. Shepard: Ecological constraints on internal representation - resonant kinematics of perceiving, imagining, thinking, and dreaming. *Psychological Rev.* **91**, 417–447 (1984)

12.26 N. Suga: Auditory neuro-ethology and speech processing - complex sound processing by combination-sensitive neurons. In *Auditory Function - Neurobiological Bases of Hearing*, ed. by G.M. Edelman, W. Gall, W. Cowan (Wiley, New York 1988)
12.27 E. Terhardt: Pitch, consonance, and harmony. *J. Acoust. Soc. Am.* **55**, 1061–1069 (1974)
12.28 N. Todd: The auditory "primal sketch" - a multiscale model of rhythmic grouping. *J. New Music Research* **23**, 25–70 (1994)
12.29 L. van Noorden: Two channel pitch perception. In *Music, Mind and Brain - the Neuropsychology of Music*, ed. by M. Clynes (Plenum Press, London 1982)
12.30 R. Zatorre: Pitch perception of complex tones and human temporal-lobe functions. *J. Acoust. Soc. Am.* **84**, 566–572 (1988)
12.31 R. Zatorre: Les excisions néocorticales temporales et leurs conséquences sur le traitement musical. In *La Musique et les Sciences Cognitives*, ed. by S. McAdams, I. Deliège (Pierre Mardaga, Bruxelles, 1989)

Chapter 13

13.1 G. Adler: *Methode der Musikgeschichte* (Breitkopf and Härtel, Leipzig 1919)
13.2 D. Baggi (ed.): *Readings in Computer Generated Music* (IEEE Computer Society Press, Los Almitos, CA 1992)
13.3 M. Balaban, K. Ebcioglu, O. Laske (eds.): *Understanding Music with AI - Perspectives on Music Cognition* (MIT Press, Cambridge, MA 1992)
13.4 A.S. Bregman: *Auditory Scene Analysis - the Perceptual Organization of Sound.* (MIT Press, Cambridge, MA 1990)
13.5 A. Camurri (ed.): Artificial Intelligence and Music. Special issue of Interface - J. New Music Research (Swets & Zeitlinger, Lisse 1990)
13.6 A. Camurri, A. Catorcini, M. Frixione, C. Innocenti, A. Massari, R. Zaccaria: Towards a cognitive model for the representation and reasoning on music and multimedia knowledge. In *Proceedings CIM 1993* (Milano 1993)
13.7 A. Camurri, M. Frixione, C. Innocenti, R. Zaccaria: A model of representation and communication of music and multimedia knowledge. In *Proceedings of the ECAI-92*, ed. by Neumann (Wiley, Chichester 1992)
13.8 C. Dahlhaus: *Untersuchungen über die Entstehung der harmonischen Tonalität (Studies on the Origin of Harmonic Tonality, transl. by R. O. Gjerdingen).* (Princeton Univ. Press, Princeton, NJ 1966/1990)
13.9 H. de la Motte-Haber: Umfang, Methode und Ziel der Systematischen Musikwissenschaft. In *Systematische Musikwissenschaft*, ed. by C. Dahlhaus, H. de la Motte-Haber (Akademische Verlagsgesellschaft Athenaion, Wiesbaden 1982)
13.10 R. Doati: Symmetry, regularity, direction, velocity. *Perspectives of New Music* **22**, 61–86 (1983)
13.11 R. Eberlein: Ein rekursives System als Ursache der Gestalt der tonalen Klangsyntax. *Systematische Musikwissenschaft* **1**, 339–351 (1993)
13.12 R. Eberlein, J.P. Fricke: *Kadenzwahrnehmung und Kadenzgeschichte - ein Beitrag zu einer Grammatik der Musik* (Peter Lang, Frankfurt am Main 1992)
13.13 J.P. Fricke: Systematische oder Systemische Musikwissenschaft. *Systematische Musikwissenschaft* **1**, 181–194 (1993)

13.14 G. Haus, I. Pighi: Stazione di lavoro musicale intelligente - l'ambiente integrato Macintosh-NeXT. In *Atti di X Colloquio Informatica Musicale* (AIMI/LIM, Milan 1993)

13.15 P. Howell, R. West, I. Cross (eds.): *Representing Musical Structure* (Academic, London 1991)

13.16 N.A. Huber: Gedanken zur Umfeld der Tonalität. *MusikTexte* **5**, 3–7 (1984)

13.17 O.E. Laske: Can we formalize and program musical knowledge? An inquiry into the focus and scope of cognitive musicology. In *Musikometrika I*, ed. by M.G. Boroda (Studienverlag Dr. N. Brockmeyer, Bochum 1988)

13.18 M. Leman (ed.): Models of Musical Communication and Cognition. Special issue of Interface - J. New Music Research (Swets & Zeitlinger, Lisse 1989)

13.19 M. Leman (ed.): Auditory models in music research (I). Special issue of J. New Music Research (Swets & Zeitlinger, Lisse 1994)

13.20 M. Leman (ed.): Auditory models in music research (II). Special issue of J. New Music Research (Swets & Zeitlinger, Lisse 1994)

13.21 C. Lischka: Understanding music cognition - an AI perspective. In *Representations of Musical Signals*, ed. by G. de Poli, A. Piccialli, C. Roads (MIT Press, Cambridge, MA 1991)

13.22 A. Marsden, A. Pople (eds.): *Computer Representations and Models in Music* (Academic, London 1991)

13.23 S. McAdams, I. Deliège (eds.): *La Musique et les Sciences Cognitives* (Pierre Mardaga, Bruxelles 1989)

13.24 G.De Poli, A. Piccialli, C. Roads (eds.): *Representations of Musical Signals* (MIT Press, Cambridge, MA 1991)

13.25 J. Risset: Hauteur et timbre des sons. Technical Report IRCAM Nr. 11 (Centre Georges Pompidou, Paris 1978)

13.26 J. Risset: Perception, environnement, musiques. *InHarmoniques* **2**, 10–42 (1988)

13.27 J. Risset: Timbre analysis by synthesis - representations, imitations, and variants for musical composition. In *Representations of Musical Signals*, ed. by G. de Poli, A. Piccialli, C. Roads (MIT Press, Cambridge, MA 1991)

13.28 J. Risset: The computer as an interface - interlacing instruments and computer sounds; real-time and delayed synthesis; digital synthesis and processing; composition and performance. *Interface - J. New Music Research* **21**, 9–20 (1992)

13.29 R. Rowe: *Interactive Music Systems* (MIT Press, Cambridge, MA 1993)

13.30 H. Sabbe: Die Sprache wiedergewonnen? Fragmente zur Tonalität im jüngsten Jahrzehnt. *MusikTexte* **14**, 30–35 (1986)

13.31 H. Sabbe: A logic of coherence and an aesthetic of contingency - European versus american "open structure" music. *Interface - J. New Music Research* **16**, 177–186 (1987)

13.32 A. Schneider: Systematische Musikwissenschaft - Traditionen, Ansätze, Aufgaben. *Systematische Musikwissenschaft* **1**, 145–180 (1993)

13.33 U. Seifert: *Systematische Musiktheorie und Kognitionswissenschaft - Zur Grundlegung der kognitiven Musikwissenschaft* (Verlag für systematische Musikwissenschaft, Bonn, 1993)

13.34 U. Seifert: Systematische musikwissenschaft als grundlagenforschung der musik. *Systematische Musikwissenschaft* **1**, 195–223 (1993)

13.35 R.N. Shepard: Circularity in judgments of relative pitch. *J. Acoust. Soc. Am.* **36**, 2346–2353 (1964)

13.36 R.N. Shepard: Structural representations of musical pitch. In *The Psychology of Music*, ed. by D. Deutsch (Academic Press, New York 1982)

13.37 C. Stumpf: *Tonpsychology I* (Hirzel, Leipzig 1883)

13.38 C. Stumpf: *Tonpsychology II* (Hirzel, Leipzig 1890)
13.39 B. Truax: Musical creativity and complexity at the threshold of the 21st century. *Interface - J. New Music Research* **21**, 29–42 (1992)
13.40 B. Vecchione: La recherche musicologique aujourd'hui - questionnements, intersciences, métamusicology. *Interface - J. New Music Research*, **21**, 281–322 (1992)
13.41 B. Vecchione, B. Bel (eds.): Music Sciences and Technologies - New Inquiries, New Connections. Special issue of Interface - J. New Music Research (Swets & Zeitlinger, Lisse, 1992)
13.42 A. Wellek: *Musikpsychologie und Musikästhetik - Grundriss der Systematischen Musikwissenschaft* (Akademische Verlagsgesellschaft, Frankfurt am Main 1963)
13.43 E. Zwicker, R. Feldtkeller: *Das Ohr als Nachrichtenempfänger* (Hirzel, Stuttgart 1967)

Subject Index

Springer Series in Information Sciences

Editors: Thomas S. Huang Teuvo Kohonen Manfred R. Schroeder
Managing Editor: H. K. V. Lotsch

1 **Content-Addressable Memories**
By T. Kohonen 2nd Edition

2 **Fast Fourier Transform and Convolution Algorithms**
By H. J. Nussbaumer 2nd Edition

3 **Pitch Determination of Speech Signals**
Algorithms and Devices By W. Hess

4 **Pattern Analysis and Understanding**
By H. Niemann 2nd Edition

5 **Image Sequence Analysis**
Editor: T. S. Huang

6 **Picture Engineering**
Editors: King-sun Fu and T. L. Kunii

7 **Number Theory in Science and Communication**
With Applications in Cryptography, Physics, Digital Information, Computing, and Self-Similarity
By M. R. Schroeder
2nd Edition

8 **Self-Organization and Associative Memory**
By T. Kohonen 3rd Edition

9 **Digital Picture Processing**
An Introduction By L. P. Yaroslavsky

10 **Probability, Statistical Optics, and Data Testing**
A Problem Solving Approach
By B. R. Frieden 2nd Edition

11 **Physical and Biological Processing of Images** Editors: O. J. Braddick and A. C. Sleigh

12 **Multiresolution Image Processing and Analysis** Editor: A. Rosenfeld

13 **VLSI for Pattern Recognition and Image Processing** Editor: King-sun Fu

14 **Mathematics of Kalman-Bucy Filtering**
By P. A. Ruymgaart and T. T. Soong
2nd Edition

15 **Fundamentals of Electronic Imaging Systems**
Some Aspects of Image Processing
By W. F. Schreiber 3rd Edition

16 **Radon and Projection Transform-Based Computer Vision**
Algorithms, A Pipeline Architecture, and Industrial Applications By J. L. C. Sanz, E. B. Hinkle, and A. K. Jain

17 **Kalman Filtering**
with Real-Time Applications
By C. K. Chui and G. Chen 2nd Edition

18 **Linear Systems and Optimal Control**
By C. K. Chui and G. Chen

19 **Harmony: A Psychoacoustical Approach**
By R. Parncutt

20 **Group-Theoretical Methods in Image Understanding**
By Ken-ichi Kanatani

21 **Linear Prediction Theory**
A Mathematical Basis for Adaptive Systems
By P. Strobach

22 **Psychoacoustics** Facts and Models
By E. Zwicker and H. Fastl

23 **Digital Image Restoration**
Editor: A. K. Katsaggelos

24 **Parallel Algorithms in Computational Science**
By D. W. Heermann and A. N. Burkitt

25 **Radar Array Processing**
Editors: S. Haykin, J. Litva, and T. J. Shepherd

26 **Signal Processing and Systems Theory**
Selected Topics
By C. K. Chui and G. Chen

27 **3D Dynamic Scene Analysis**
A Stereo Based Approach
By Z. Zhang and O. Faugeras

28 **Theory of Reconstruction from Image Motion**
By S. Maybank

29 **Motion and Structure from Image Sequences**
By J. Weng, T. S. Huang, and N. Ahuja

Springer-Verlag and the Environment

Printing: Saladruck, Berlin
Binding: Buchbinderei Lüderitz & Bauer, Berlin